D1399067

Integrated Circuits

Integrated Circuits

A Basic Course for Engineers and Technicians

R. G. Hibberd
Texas Instruments Limited
Bedford, England

, LIBRARY OF
JOHN D. ANDERSON

McGRAW-HILL BOOK COMPANY

New York   St. Louis   San Francisco   London
Sydney   Toronto   Mexico   Panama

53657

INTEGRATED CIRCUITS

Copyright © 1969 by Texas Instruments Incorporated. All Rights Reserved. Printed in the United States of America. No part of this publication may be reproduced, stored in a retrieval system, or transmitted, in any form or by any means, electronic, mechanical, photocopying, recording, or otherwise, without the prior written permission of Texas Instruments Incorporated. *Library of Congress Catalog Card Number* 69-18728

ISBN 07-028651-5

890 HDBP 754

Information contained in this book is believed to be accurate and reliable. However, responsibility is assumed neither for its use or infringement of patents or rights of others which may result from its use. No license is granted by implication or otherwise under any patent right of Texas Instruments or others.

To my wife, Dora,
my sons, Stuart, Robert, and Bernard,
and my daughters, Kay and Carole

Preface

The development of solid-state technology has resulted in the application of electronics to more and more aspects of our everyday life. With the evolution of integrated circuits, the use and application of electronics is increasing at a considerable rate, even in areas that were previously considered the property of the mechanical or industrial engineer rather than the electronics engineer. The integrated circuit—a complete, functional electronic circuit fabricated in a tiny wafer of silicon—now enables the user, whether he be an electronics engineer, an electrical engineer, a mechanical engineer, or a general industrial engineer, to be less concerned with the details of the internal circuitry than with its overall function, defined by certain input and output parameters.

It is becoming more and more essential for every engineer to have a general, basic understanding of solid-state technology and integrated circuits so that he can better evaluate the potential use of integrated circuits in his products. He needs to be generally informed of the present state of the art and future trends so that he can be in a position to consider alternative approaches to any particular application, from the viewpoint of what is available today and also what will become available within the near future. Then he can minimize the possibility of new advances rendering his developments obsolete before they are in full manufacture.

We cannot all be experts in every subject, but it is essential that the systems engineer and the mechanical engineer at least be able "to talk the same language" as the integrated-circuit experts. Extensive interaction between the users and designers of integrated circuits is necessary in order to convert the potential uses of integrated circuits into the most efficient and profitable products.

This course is aimed at giving the nonelectronics expert a general understanding of the fundamentals and application of integrated circuits. A previous course on "Solid-state Electronics" (McGraw-Hill) is recommended as introductory reading.

R. G. Hibberd

Contents

The Impact of Integrated Circuits

INTRODUCTION

The transistor was a radically new innovation in electronics. Before it could be universally accepted as the new basis for electronic circuitry, it was necessary to evolve new circuit theory, to develop new engineering and technical processes, and to establish new manufacturing facilities for large-volume production. As a result, it was between ten and fifteen years before solid-state electronics became firmly established, and so the impact of the transistor, while revolutionary in nature, was relatively gradual in effect.

The silicon monolithic integrated circuit is a further development of the research and manufacturing activities established for the transistor. In general, integrated circuits are fabricated with the same materials and processes as those used in the manufacture of the epitaxial planar transistor, and so the experience built up on the transistor was directly applicable to integrated circuits. This resulted in a shorter period of time—only about seven years—for integrated circuits to be developed and accepted by the electronics industry.

There can be no question of the widespread influence of integrated circuits on electronics. There has been a steady increase in their application, for example, in missiles and computers where their particular assets of small size and high reliability offered immediate advantage and, more recently, in industrial and consumer equipment in which low cost is the attractive feature.

In looking at the overall impact of integrated circuits, several aspects must be considered. Perhaps the greatest effect will be on the customer or user of integrated circuits—the electronic equipment manufacturer. Traditionally, he has procured discrete components and then carried out circuit design and fabrication before building the complete equipment. Now, he has available to him at a lower cost per circuit function, complete circuits which are smaller and more reliable, with at least comparable, and in many instances, better, performance. To be economically competitive, he must use them: this means that he has less circuit design to do and he can devote more effort to improving the operation and versatility of the complete equipment.

The integrated-circuit manufacturer, on his part, must establish a comprehensive circuit-design capability and interface to a greater extent with the equipment manu-facturers to ensure that their performance requirements are met in the most eco-nomical manner. The detailed design of integrated circuits must involve a judicious combination of the required performance with what is achievable with the tech-nological processes involved in their fabrication.

The third and longer-term impact of integrated circuits will be on the whole structure of the electronics industry. Because he is now fabricating complete circuits and as technology improves he will be able to include more and more circuit functions on a single chip of silicon, the integrated-circuit manufacturer will grad-ually move toward supplying complete subsystems, and he will engage in an even bigger part of the industry. In general, electronics engineers will need to have a wider technical background covering from physics through electronic circuitry to equipment operation to enable them to interface skillfully, both at the intake to their specific sphere of operation and also at their output.

In this opening lesson, the various aspects contributing to the general impact of integrated circuits will be discussed.

IMPLICATIONS OF SMALL SIZE

In the monolithic integrated circuit, all the components of the circuit, both active and passive, are formed at the same time within a small wafer of silicon typically between $\frac{1}{20}$ and $\frac{1}{10}$ in. square. The small size of the integrated-circuit wafer means that the final encapsulated unit can be correspondingly small, and the first standard package was only about $\frac{1}{4}$ in. long by a little more than $\frac{1}{8}$ in. wide. So it is not surprising that the first impact of the integrated circuit was the great reduction in size of electronic equipment that would be possible. Considerable impetus for the development of early types came from the space program. In space vehicles, every pound of weight saved means a reduction in space-vehicle cost of about $20,000, due to the decreased requirement for fuel to give the thrust to place the vehicle into orbit. The integrated circuit promised considerable savings in this respect. There was additional impetus from the military to reduce the weight and size of airborne electronic and portable communication equipment, which was continually becoming more complex and bulky.

As experience with the use of integrated circuits built up, it soon became apparent that other considerations—namely, the possibility of higher equipment reliability and lower cost—were even more significant, and these considerations, rather than the small size, became the prime objectives of integrated-circuit development. This is not to say that the small size was not considered important. As will be discussed later, by making the integrated-circuit wafer as small as possible, the processing cost per wafer decreases, and the fabrication yield increases, both contributing to lower cost. But apart from space applications, it appeared that there was not the need to encapsulate this small wafer into the very small package. For much ground-based equipment, the overall size of the system is often dictated by other components or peripheral equipment. A good example of this is a TV receiver, where the size is largely dictated by the picture tube. As a result, larger, low-cost packages were developed to be compatible with existing printed-circuit-board assembly techniques.

Since the dimensions of the packaged unit are an important concern of the user, it may be well to review the alternative versions. There are three standard packages, illustrated in Fig. 1.1. For applications where small size is most important, a small package called a *flat-pack* is used (Fig. 1.1c). This has overall dimensions, excluding leads, of 0.25 in. long, 0.15 in. wide, and 0.05 in. thick. It is a hermetically sealed unit with glass-to-metal seals for the leads. The second package is a multilead version of the transistor TO-5 package (Fig. 1.1a). The larger alternative encapsulation (Fig. 1.1b) is called the *dual-in-line* package. The leads are on 0.1-in. centers so that they align with mounting holes in a standard printed-circuit board. The dimensions of this package are 0.77 in. long, 0.25 in. wide, and 0.2 in. thick. The illustration shows a molded plastic construction which is satisfactory for most industrial applications. A sealed ceramic unit with similar dimensions is available when a fully hermetic unit is required.

Another aspect of small size concerns the circuit operation. The close spacing of the several elements and interconnections within the unit reduces the possibility of unwanted electrical pickup from fluctuating magnetic fields, and so integrated circuits are able to operate at lower signal levels. This allows the use of lower supply voltages with resulting lower operating-power requirements. Although this may not be great for one circuit, the cumulative saving for large equipments can be very significant.

DESIGN CONSIDERATIONS OF INTEGRATED CIRCUITS

The technological processes involved in the fabrication of monolithic integrated circuits impose certain restraints which must be carefully considered at the design

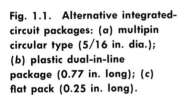

Fig. 1.1. Alternative integrated-circuit packages: (a) multipin circular type (5/16 in. dia.); (b) plastic dual-in-line package (0.77 in. long); (c) flat pack (0.25 in. long).

stage. These restraints include aspects relating to relative cost and performance, and so may affect the optimum design of any particular unit.

In earlier electronic circuits using discrete components, the active devices, vacuum tubes and transistors, were generally the most expensive components. The passive components, particularly the resistors, were relatively inexpensive. Thus, to obtain low cost, the circuit design was aimed at achieving the desired result with the minimum number of active devices.

With integrated circuits, the cost of processing a silicon slice through the several oxide-removal, diffusion, and metallization stages is roughly independent of the content of the slice. The smaller the area of a circuit, the greater the number of circuits on the slice and the lower the processing cost per circuit. Thus, the cost of processing a circuit is roughly proportional to the total area it occupies on the silicon slice, and the same consideration applies to each element within the circuit. Figure 1.2 shows the relationship between area and element value for diffused resistors, junction capacitors, and oxide capacitors, as formed in integrated circuits. Included are lines showing the area occupied by a typical integrated-circuit transistor and diode. It will be seen that above about 1,000 ohms and 10 pf, the resistors and capacitors take up more area than a transistor and so cost more. As a result, integrated circuits are designed to use transistors and diodes wherever possible, rather than resistors and capacitors, in order to reduce the total area and so the fabrication

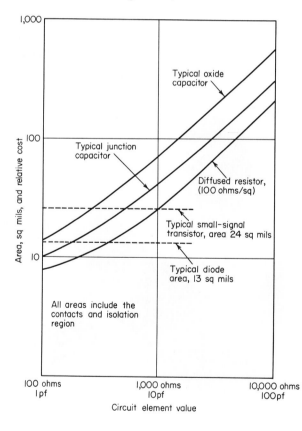

Fig. 1.2. Relations between area and value of integrated resistors and capacitors.

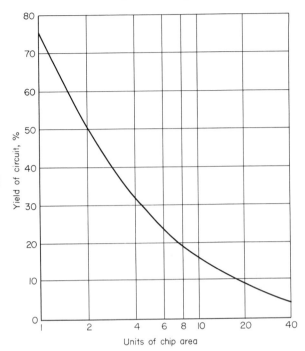

Fig. 1.3. Effect of chip area on the yield of an integrated circuit.

cost to a minimum. Direct coupling between stages to avoid capacitors and low-value resistors will offer the lowest-cost circuits. It may be possible to use a transistor as an adjustable resistor and effect a saving. The necessity to isolate the various elements from the substrate increases the area they occupy, and the design must explore the possibility of including more than one element in a common isolation region to reduce the total area.

There is another consideration which makes small circuit area even more important. In practice, it has been found that the yield of good circuits from a slice increases as the area of the circuit is made smaller. This is illustrated in Fig. 1.3 and is almost certainly due to random imperfections which are either present in the original silicon material or which develop in it during the high-temperature slice processing. The larger the circuit area, the greater the possibility of a defect being within the area and adversely affecting the circuit. Thus the motivation to design integrated circuits to have as small an area as possible is very great, and the impact of this has been to move the fabrication of integrated circuits into a precision industry, with element dimensions down to a few tenths of a mil with tolerances an order of magnitude smaller. This has required the development of much precision manufacturing equipment, and several companies have come into existence just to make and provide such equipment to integrated-circuit manufacturers.

As the capability of greater precision develops, the integrated-circuit manufacturer can make any particular circuit in a smaller area with consequent lower cost, or alternatively, he can put more elements into the same area and produce more complex circuits at the same cost.

There are certain features of the circuit elements in an integrated circuit that are not present with discrete-component circuits and necessitate careful attention during design to ensure that unwanted effects will not occur. The isolation junction has a capacitance which could allow parasitic coupling at high frequency. The diffused resistors have distributed capacitance to the region in which they are formed, and this may lead to a change in effective resistance at high frequency. The value of a junction capacitor depends on the bias voltage across it which may require special arrangement in the circuit design. At the present time, it is not possible to reproduce the absolute value of a diffused resistor better than about ± 5 percent. As a result, circuit design is often arranged so that the operation relies on the ratio of two resistors, as they can be fabricated side by side with a very close tolerance, about ± 1 percent, on the ratio of their resistances. This is of importance in negative-feedback circuits.

PERFORMANCE

In the design of integrated circuits, it may be necessary to compromise in some aspects of performance since all elements are formed by the same sequence of diffusions. It is normal for the diffusion processes to be optimized for the main transistors in the circuit, and this may not be ideal for other transistors and elements in the circuit. For example, it may be desirable to make some of the transistors have a fast switching action. Since all the other elements are formed by the same process, they will have similar characteristics, and this may impose a restriction which would not be present with discrete-device circuits, in which each component can be independently specified. A reasonable compromise is usually possible, however, and with careful design, integrated circuits can give performance similar to the corresponding discrete-device circuit.

It might appear that there is some limitation in allowable power dissipation with the very small integrated-circuit wafers, but for low-level operation there is adequate power capability—up to about 250 mw per circuit. There is no fundamental reason why low-level control circuits cannot be integrated on the same wafer with power-device structures, and then the thermal design of the wafer mounting system will look after the required level of power dissipation.

One instance where the integrated circuit offers improved performance is where thermal "tracking" of devices is desirable, such as with the pair of input transistors in a differential dc amplifier. Here it is required to balance out the dc drift due to temperature change. Two transistors fabricated side by side in a silicon wafer will have substantially identical characteristics and will track closely over a wide range of temperature.

Another possible improvement is that the integrated circuit may allow shorter total switching time. With a discrete-device logic circuit, it is not possible to use the full speed capability of the transistors because of stray capacitance and propagation delay associated with the wiring between components. The small size and close proximity of elements in integrated circuits give the possibility of reducing the stray capacitance and the time delay. Also, because the wiring between circuits can be shorter, a further reduction in overall system delay time is possible.

An area where the integrated-circuit approach offers much improved performance is the microwave region. Microwave integrated circuits will be truly *micro* from both the operating-wavelength and the physical-size viewpoint. An important point is the compatibility of dimensions between microstrip lines and circuit-element dimensions. For example, if a high-resistivity silicon wafer 10 mils thick is used as the dielectric of a microstrip transmission line, a characteristic impedance of 50 ohms will be obtained with the top conductor only 5 mils wide. This is reasonably similar to device dimensions and allows good interfacing between devices and strip lines. Because of certain difficulties concerned with maintaining the resistivity of silicon at a high value through the high-temperature processes, microwave integrated circuits are presently of the hybrid type, using ceramic substrates with thin-film strip-line conductors and silicon transistors and diodes in chip form, but the general comments above still apply.

RELIABILITY

The ever-increasing complexity of modern electronic equipment, requiring large numbers of individual components connected together to form an overall system, has brought great emphasis on higher reliability.

Reliability can be defined as "the probability that a device or system will perform a specified function for a specified period of time without failure." To look further into what this means, consider a modern electronic computer which may contain 50,000 transistors and diodes. The failure of one of these may render the computer inoperative, and if it is specified that the computer must operate for an average of 7 days without a component failure, the devices must each have a life expectancy of 350,000 days, or approximately 1,000 years. Since we cannot monitor or specify such a period of time, it has become conventional to assume a constant failure rate with time and then to specify a maximum failure rate for the devices. In the above case, this will be a maximum failure rate of 1 device out of 50,000 in 7 days (168 hr). Stated in another way, this is a failure rate of 0.012 percent per 1,000 hr. This is the way that failure rates are normally quoted for integrated circuits.

It has been discussed earlier how the processes used in the fabrication of integrated circuits are basically the same as those used for silicon planar transistors. In view of this, it could be argued that the reliability of an integrated circuit should be of the same order as that of a silicon planar transistor of equal wafer size. Results to date have indicated that this may well be a reasonable argument, and it does appear that the reliability of an integrated circuit is generally independent of the number of elements in a given wafer size. This is a most important factor. Recently designed types of integrated circuits can include up to 50 elements using a wafer size previously used for a single planar transistor. This indicates the possibility of an improvement in equipment reliability up to the order of 50 times.

The failure rate of present-day integrated circuits is typically better than 0.01 percent per 1,000 hr, and reliability data from equipment using large numbers of integrated circuits in the field is indicating a figure approaching 0.001 percent per 1,000 hr.

The measurement of this order of reliability by the integrated-circuit manufacturer represents no small problem. It is necessary to put large numbers of units on life test under specified operating conditions, for long periods of time. The failure rate is then quoted as the number of failures divided by the product of the number of devices on test and the total number of operating hours. For example, if 10,000 devices are put on test for 1,000 hr (6 weeks) and there is 1 failure, the failure rate is 1 per 10,000,000 device hr, which is 0.01 percent per 1,000 hr.

The overall reliability of a complete electronic equipment is often quoted as the *mean time before failure* (MTBF). This is defined as the average time between successive failures of the equipment. It can be estimated as the reciprocal of the sum of the failure rates of each individual component and connection in the system. Systems using integrated circuits have shown MTBF improvements up to 50 times when compared with comparable equipment using discrete-component assemblies.

This short discussion will have indicated that integrated circuits are making possible a much higher system reliability than was possible with discrete-component assemblies. For applications such as missiles, when the system must operate 100 percent on demand, and the cost of a single operational failure can amount to millions of dollars, the high reliability of integrated circuits is probably the most important reason for using them. The small size and low cost in these applications constitute a very welcome bonus.

OVERALL COST CONSIDERATIONS

It is interesting to look at the overall cost of an integrated circuit. The design cost will be quite high, since it includes the initial functional circuit design, often in "breadboard" form, then the design of each individual circuit element and the geometrical layout including the interconnection pattern, the design of the several (at least six) photomasks for the oxide-removal stages, the processing of trial proving batches, and the development of suitable test programs. The design having been proved, however, the straight manufacturing cost will be quite small. Compared with that for a silicon planar transistor, the manufacturing process includes two or three more photoresist stages, and the assembly and test cost will be somewhat higher. Considering these points, an integrated circuit should not basically cost more than about twice the cost of a planar transistor of the same wafer area. At present it costs somewhat more than this because we are not as far down the "learning curve" with integrated circuits, and a figure between three and four times the transistor cost is perhaps more realistic. If the design cost is allocated pro rata over the total number of circuits manufactured, the overall cost per unit will depend on the total number manufactured as shown in Fig. 1.4. A typical integrated-circuit design cost is taken as $10,000 and the manufacturing cost as $1.50. The overall cost is the sum of the prorated design cost (curve *A*) and the manufacturing cost (curve *B*). With these figures it will be seen that the allocated design cost per unit comes down to 10 percent of the manufacturing cost when about 60,000 units are made. Below that number, the overall cost per unit increases and becomes very dependent on the number manufactured, and above, the overall cost approximates to the manu-facturing cost.

Included in Fig. 1.4 is an estimate of the corresponding costs for a discrete-device printed-circuit-board assembly of the same circuit content as that of the integrated

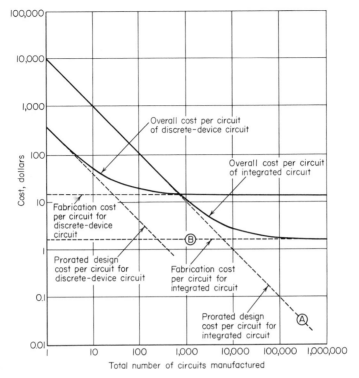

Fig. 1.4. Relation between the overall cost per circuit and the total number of circuits manufactured.

circuit, i.e., functionally equivalent. It can be inferred from Fig. 1.4 that the overall cost of an integrated circuit can show significant advantage over the cost of discrete-device circuits, *but only if large numbers are manufactured.* If a user wishes to obtain a custom-designed integrated circuit but only requires, say, 100 to be manufactured, with the costs as in the above example, the total cost per unit will be $100, which is uneconomically high when compared with the discrete-device assembly. If he could accept a standard circuit, which the integrated-circuit manufacturer is making in high volume and selling to many users as a catalog item, the cost would be back down to $1.50. Thus it appears that there will be a growing acceptance of standard integrated circuits by the equipment industry in order to obtain low-cost units.

In addition to offering the low cost in itself, the integrated circuit allows further cost reduction in electronic equipment manufacture, since a smaller number of separate units have to be assembled for any given equipment. There will also be a reduction in the overhead costs of procurement and inventory.

EFFECT ON THE ELECTRONICS INDUSTRY

The fact that the integrated circuit is manufactured as a single component by component manufacturers but is a complete functional circuit that previously would have been designed and manufactured with discrete components by the equipment engineer means that there must be a redistribution of activities within the electronics industry.

With discrete components, the electronics industry was traditionally organized as shown in the first column of Fig. 1.5, with two groups of manufacturers making and supplying components, a third group designing circuits and making complete electronic equipments incorporating the circuits, and a fourth group manufacturing large systems. The latter activity is often carried out by the equipment group. The semiconductor-component manufacturer, in making integrated circuits, is moving into the equipment activity, and as the complexity of integrated circuits increases toward what are called integrated electronic components, he will move in even further. As a result, the division of activities is moving as indicated by the heavy dashed lines in Fig. 1.5, with the semiconductor-component manufacturer gradually operating in a larger part of the industry. With discrete-device circuits, the equipment manufacturer tended to trade his equipment on the uniqueness of his particular circuit designs. With integrated circuits he is no longer designing his own circuits; although the equipment manufacturer may contribute to the design by discussion and specification, in most cases, the integrated-circuit manufacturer will actually design the circuits and sell them as standard circuits to any equipment manufacturer. So the equipment manufacturer must now introduce uniqueness into his equipment by combining standard circuits to give more comprehensive operation with better reliability at a lower cost.

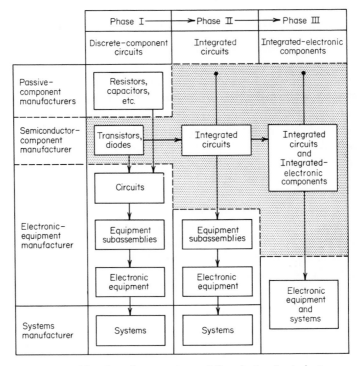

Fig. 1.5. The changing structure of the electronics industry.

The use of integrated circuits influences the activities of service and maintenance of electronic equipment. With the higher system reliability resulting from the use of integrated circuits, there will be less field maintenance required to keep large installations in an operating condition. However, it must be remembered that an integrated circuit cannot be repaired, even though only one circuit element may have failed—the whole circuit must be replaced. At first sight, it might appear that this would make the repair very costly, but there are compensating factors. It is easier to diagnose a faulty circuit than a faulty component, and so the need for highly skilled technicians is reduced, and a smaller number of spare parts will be required, reducing inventory costs.

Finally, integrated circuits demand integrated thinking. New technical language and new disciplines must be mastered by the component manufacturer, the equipment manufacturer, and the systems manufacturer. Each must learn more about the others' operations so that they can interface more to foster the integrated thinking. Electronics engineers now need to have a wide technical knowledge, and managers an equally wide background.

CONCLUSIONS

The integrated circuit has already made a significant impact on the electronics industry, and it will continue to become more and more important. It has been estimated that 70 percent of all existing electronic circuits could be produced in integrated form with the attendant advantages of higher reliability and lower cost.

This, however, is not the whole story. Electronics can often be a better alternative way of doing something that has previously been done mechanically or electromechanically, such as power control, switching, counting, and power conversion. Integrated circuits will add impetus to this trend. With integrated circuits being specified on a functional basis, it will be possible for the mechanical engineer to plan their use in conjunction with the mechanical parts of his equipment without having to understand the detailed way in which the function is performed.

But perhaps the most far-reaching application of integrated circuits is in electronic computers, which are now being used for a wide range of activities which will have a considerable impact on all ways of our everyday life.

GLOSSARY

active device   A device which displays gain or control, such as a transistor or vacuum tube.

failure rate   The rate at which a particular type of device will fail. It is usually expressed as the percentage of units failing in a time of 1,000 hr.

manufacturing cost   The cost of manufacturing a device once the design has been completely specified and the manufacturing facilities are set up.

monolithic integrated circuit   A complete electronic circuit which has been fabricated as an inseparable assembly of circuit elements in a single small structure which cannot be divided without permanently destroying its intended electronic function.

MTBF (mean time before failure)   The average time between successive failures of a system. It is the reciprocal of the sum of the failure rates of every component and connection in the system.

overall cost   The total cost of manufacture plus the total cost of design (both including all overheads) divided by the total number of devices manufactured.

passive device   A device not displaying gain or control, e.g., a resistor or a capacitor.

reliability   The probability that a device or system will satisfactorily perform a specified function for a specified length of time.

transistor   A solid-state device, with three or more electrodes, capable of giving power amplification.

yield   The ratio, expressed as a percentage, of the number of good devices produced to the maximum possible number if all were good.

REVIEW

For each of the numbered statements below, select the one of the items lettered *a, b, c,* or *d* that correctly completes the statement.

1.1. Monolithic integrated-circuit design is based on
   a. Extensive use of resistor-capacitor coupling.
   b. Using high values of resistors and capacitors.
   c. Using transistors and diodes wherever possible, rather than resistors and capacitors.
   d. Making the area of the circuit elements as large as possible.

1.2. The cost of fabricating a monolithic integrated circuit is
   a. Proportional to the number of circuit elements.
   b. Roughly proportional to the area of the circuit wafer.
   c. Independent of the number of passive elements.
   d. Independent of the yield.

1.3. The yield of good integrated circuits on a silicon slice
   a. Is independent of the quality of the original silicon slice.
   b. Is always the same.
   c. Does not depend on the number of defects in the original silicon slice.
   d. Increases with decrease of circuit area.

1.4. The small weight and size of integrated circuits
   a. Makes them very attractive for use in missiles.
   b. Is always their most important advantage.
   c. Is most important in consumer applications.
   d. Is a disadvantage in circuit operation.

1.5. Monolithic integrated circuits can give better performance than the corresponding discrete-component circuit
   a. In no application.
   b. In any application.
   c. In differential dc circuits where thermal tracking is necessary.
   d. Only in some switching circuits.

1.6. Microwave integrated circuits
   a. Are not possible.
   b. Will have inferior performance.
   c. Are initially being made as hybrid types on ceramic substrates.
   d. Will always use discrete components.

1.7. Advantages of integrated circuits include the possibility of
   a. Repairing individual circuit elements.
   b. Obtaining high-tolerance resistors.
   c. Using high values of capacitors.
   d. Obtaining high reliability at low cost.

1.8. The overall cost of an integrated circuit
  a. Is always lower than the corresponding discrete-component assembly.
  b. Depends upon the total number manufactured.
  c. Is always dominated by the design cost.
  d. Is tending to increase.
1.9. The reliability of a monolithic integrated circuit
  a. Is easy to measure.
  b. Results in a short MTBF.
  c. Is similar to that of a single silicon planar transistor with the same wafer size.
  d. Is worse than discrete-component assemblies.
1.10. The effect of the integrated circuit on the electronics industry will be
  a. A redistribution of activities within the industry.
  b. Insignificant.
  c. The establishment of a greater maintenance and repair activity.
  d. An increase in the types of companies.

LESSON 2

Solid-state Technology

INTRODUCTION

A solid-state device is an electronic device which operates by virtue of the movement of electrons within a solid piece of semiconductor material. The use of solid-state devices in electronic equipment is not new; some devices are as old as the electronics industry itself. For example, a semiconductor galena crystal was used as a detector of radio signals in the very early experiments on radio communication. For many years, however, semiconductors played a minor role in electronics because of the development and use of the vacuum tube. The present emphasis on the solid-state approach was initiated by the invention of the transistor in 1948 by W. Shockley, J. Bardeen, and W. H. Brattain. Starting with its early application in pocket radio receivers and hearing aids, the transistor has been responsible for a complete revolution in most fields of electronics, particularly in the design of computers and satellites. In addition, a whole new family of related solid-state devices has been developed—field-effect transistors, diodes, power rectifiers, controlled rectifiers, photodiodes, and many others—culminating in the development of monolithic integrated circuits, complete electronic circuits formed in a tiny piece of silicon material.

The word *tiny* used in the previous sentence is truly descriptive. A typical integrated-circuit wafer measures $\frac{1}{20}$ in. square by 10 thousandths in. (mils) thick and contains about 50 electronic components. Typical dimensions for the components in such a circuit are: transistors, 6.5×4 mils; diodes, 4.5×3 mils; and resistors, 2×12 mils. These components are formed close together in the wafer of silicon and interconnected to form the required electronic circuit by a metal pattern evaporated onto the top surface. The wafer is assembled and sealed into a container having the requisite number of leads, giving a unit approximately $0.26 \times 0.15 \times 0.040$ in. thick.

This small size is only part of the revolution brought about by solid-state technology. The potential capabilities and properties of integrated circuits are so wide that, in addition to replacing similar discrete-component circuits, they will result in a completely new technology of circuit and system design.

To be able to realize the full implication of integrated circuits, it is essential to

14

have a general, basic understanding of solid-state technology and its application. After a general introduction covering the basic principles of solid-state semiconductor devices, this lesson will review the processes involved in the formation of solid-state structures that are used in integrated circuits.

BASIC PRINCIPLES OF SOLID-STATE DEVICES

In a *conductor,* a flow of electric current consists of a movement of free electrons. The outer or valence electrons of a good conductor such as copper are so loosely bound to the atom that at room temperature the thermal energy causes approximately one electron to become detached from each atom to move freely and result in a current flow when an electric potential is applied.

Insulators are materials in which the outer electrons are tightly bound to the atom, and no electrons are free to move. Thus, no current can flow when a voltage is applied.

Between these two major categories is a class of materials called *semiconductors.* As the name implies, a semiconductor is a material with conductivity roughly midway between conductors and insulators. However, a semiconductor is not just a poor conductor; it has two other very important properties. First, its resistance normally decreases with increase of temperature, as opposed to conductors such as metals with which the resistance increases slightly with temperature. Second, flow of current in a semiconductor can be by two mechanisms: either by a flow of negative electrons similar to current flow in conductors or by a movement of missing electron sites in the opposite direction. If an atom has one outer electron missing, a loosely bound electron from a neighboring atom can jump into it, leaving behind a new vacant site; this in turn can be filled by an electron from a third atom, and so on. It then appears that the vacant site has moved. Such vacant sites are called *holes* and, since a negative electron is missing, can be considered positive charges. This concept of hole movement to give a current flow is very important in the understanding of semiconductors.

Semiconductor material in which conduction is by a flow of electrons is called n-type material (n for *n*egative carriers), and material in which conduction is due to the movement of positive holes is called p-type (p for *p*ositive carriers).

From the viewpoint of monolithic integrated circuits, the most important semiconductor material is silicon. Silicon has four outer or valence electrons. If we add to it a small amount of an impurity element with five valence electrons, such as phosphorus, one electron per impurity atom will be free, and we will have *n-type silicon.* Similarly, if we add an impurity with only three valence electrons such as boron, there will be one missing electron or hole per boron atom, and we will have *p-type silicon.* Thus, we can readily control the type of conduction in silicon material by doping it with a suitable impurity.

The p-n Junction. The operation of most solid-state devices is dependent on the properties of one or more p-n junctions incorporated into their structure. A p-n junction is a transition from a p-type semiconductor to an n-type semiconductor within a piece of material. Alone, a piece of n-type or p-type semiconductor is purely resistive; that is, reversing a battery connected across it will reverse the direction of current flow but will not affect the magnitude of the current. In contrast, a piece of semiconductor material with a p-n junction in it has rectifying properties. When the positive terminal of a battery is connected to the p-type side and the negative terminal to the n-type side, the free negative electrons in the n-type side are attracted across the junction to the positive contact, and the positive holes in the p-type side are attracted across the junction in the opposite direction to the negative contact, as illustrated in Fig. 2.1a. This is called the *forward* or *conducting direction*. A high current flows with only a small applied voltage, and the forward resistance is very low. Now if the battery connections are reversed, as in Fig. 2.1b, the holes in the p-type side are attracted away from the junction toward the negative terminal, and the electrons in the n-type side are attracted away from the junction to the positive terminal, and so no current flows across the junction. This is called the *reverse* or *nonconducting direction,* and no current flows even with a high voltage applied. (In practice a very small leakage current flows, due to free electrons and holes being generated near the junction by the thermal energy.) Thus it can be seen that a single p-n junction can be used as a rectifying diode for many circuit functions.

An important point to observe is that when forward current is flowing through a p-n junction as in Fig. 2.1a, electrons are flowing through p-type material in which there are normally no free electrons and some holes are flowing through n-type material. The electrons are said to have been injected across the p-n junction into the p-type material. It is this situation that leads to the operation of the junction transistor.

The n-p-n Junction Transistor. A junction transistor consists of two p-n junctions formed in a piece of semiconductor material with a very small separation, only

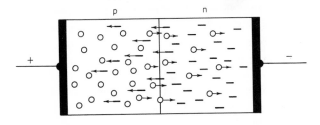

(a) With forward bias

— Free electrons
o Positive holes

Fig. 2.1. Operation of the p-n junction.

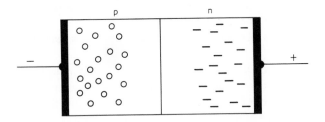

(b) With reverse bias

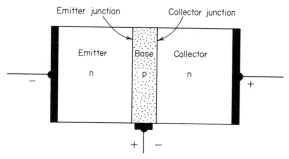

(a) Diagrammatic representation

Fig. 2.2. Basic n-p-n junction transistor details.

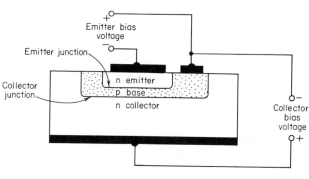

(b) Practical arrangement

of the order of 0.1 mil. Figure 2.2 shows a diagrammatic n-p-n transistor with its operating voltages. The first n-type region is called the *emitter,* since it emits or injects electrons into the center p-type region which is called the *base.* The second n-type region is called the *collector,* as it collects electrons from the base region. The junction between emitter and base is called the *emitter junction,* and that between collector and base the *collector junction.* As mentioned above, in practice, the width of the base region is only of the order of 0.1 mil.

The collector junction is biased with a high voltage in the reverse direction (positive to the n-type collector, negative to p-type base), and so, considering the collector junction by itself, no current flows across it. Now suppose the emitter junction is biased in the forward direction (negative to the n-type emitter and positive to the p-type base). A forward current flows across the emitter junction, and electrons will be injected into the p-type base region. If there were no voltage applied to the collector junction, these electrons would flow out of the base contact; but with the collector reverse-biased, as soon as the electrons in the base region flow near the collector junction, they are attracted across it by the positive potential on the collector side. Thus most of the current crossing the emitter junction continues on across the collector junction. The current across the emitter junction was produced by a very low forward voltage (less than 1 volt), and this current now flows in the collector circuit which is biased with a high voltage, and we have power amplification.

As the electrons injected across the emitter junction flow through the base region,

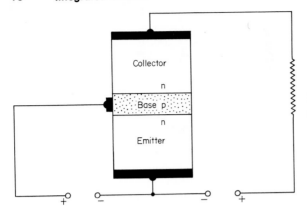

Fig. 2.3.   Common-emitter circuit arrangement.

some of them will fill holes existing in the p-type material. Thus the electron current crossing the collector junction will be slightly less than the emitter current. The ratio of collector current divided by emitter current (I_C/I_E) is called the current transfer ratio and is designated by the symbol α; that is, $\alpha = I_C/I_E$. A typical value for α is 0.98.

A current equal to the difference between the emitter current and the collector current flows in the base lead so that

$$I_E = I_C + I_B$$

With $\alpha = 0.98$, the base current I_B will only be 0.02 times the emitter current.

In the description of transistor action described above, the current through the collector is controlled by the emitter current, and the arrangement is called the *common-base circuit*, since the base electrode is common to both input and output circuits. A more convenient arrangement is with the input fed to the base and the emitter common to both input and output, as shown in Fig. 2.3. This is called the *common-emitter circuit*. The collector current is now effectively controlled by the base current I_B, and the ratio of collector current to base current has a high value, equal to $\alpha/(1 - \alpha)$. If $\alpha = 0.98$ as above,

$$\frac{I_C}{I_B} = \frac{\alpha}{1 - \alpha} = 49$$

Thus we have a current gain from the input to the collector of 49 times, and if the collector current flows through a load resistance to give a voltage output from the collector, a voltage gain results.

The p-n-p Junction Transistor.   A p-n-p structure will operate in a similar way to the n-p-n transistor described above. There are two points to observe. First, in order to bias the emitter junction in the forward direction, the emitter must be made positive with respect to the base, and to bias the collector junction in the reverse direction, the collector must be made negative with respect to the base. Second, the p-type emitter will inject positive holes into the n-type base region, and they will subsequently be attracted across the collector junction by the negative potential in the reverse direction.

CIRCUIT SYMBOLS

In order to represent the transistor graphically, the symbols shown in Fig. 2.4a are used. The emitter is shown as an arrow pointing in the direction of positive current flow (opposite to the direction of electron flow).

Figure 2.4b shows the arrangement of Fig. 2.3 in the symbolic form.

SOLID-STATE TECHNOLOGY

The above discussions have indicated the importance of the p-n junction. The technology of fabricating solid-state devices and integrated circuits has evolved around the development of methods of producing p-n junctions in silicon material. The following sections will be devoted to a general description of the basic processes involved in solid-state technology.

Silicon Preparation.   Silicon is a metallic element with a light gray appearance. It occurs in nature as silicon dioxide (silica) and as various silicate compounds. To prepare silicon for solid-state devices, there are two main requirements. First, extremely high purity is required, with unwanted impurities down to a level of 1 part in 10^{10}. Second, for a p-n junction to operate as described earlier, the silicon must have a continuous regular crystal structure, and so the silicon must be converted into what is called *single-crystal form.*

The first step in the preparation of semiconductor-grade silicon is to reduce silica by heating it with carbon (coke) in an electric furnace. The resulting silicon is about 98 percent pure. The next step is to purify this material. It is converted to a compound such as a halide (silicon tetrachloride) which is purified by repeated distillation. Then the purified halide is converted back to silicon by hydrogen reduction. In this process, the silicon is deposited onto the surface of a high-purity silicon rod, building it up to a diameter between 1 and 4 in. Using this chemical method of purification, the required purity level of 1 part in 10^{10} can be achieved, and the resulting silicon is suitable for the preparation of solid-state devices and integrated circuits.

The silicon deposits onto the rod in polycrystalline form, and this must now be converted to single-crystal form. The process generally used to produce single-crystal silicon for transistors and integrated circuits is called *crystal pulling.* Figure 2.5 shows the general arrangement. Solid polycrystalline silicon is placed in a pure quartz

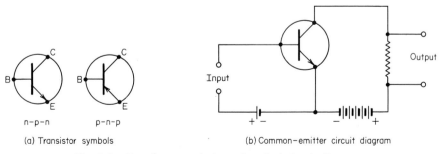

(a) Transistor symbols   (b) Common–emitter circuit diagram

Fig. 2.4. Transistor symbols and typical circuit diagram.

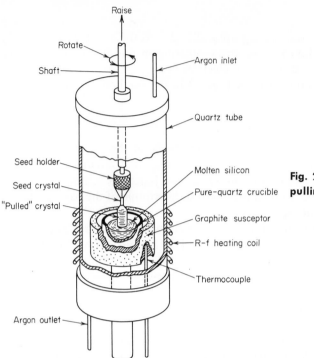

Fig. 2.5. Silicon crystal-pulling furnace.

crucible supported inside a transparent quartz chamber, through which a flow of an inert gas such as argon is maintained. The quartz crucible is located in a graphite susceptor, which is heated by r-f induction. When the silicon is molten, its temperature is lowered to a value just above its melting point, and a seed crystal—a small piece of single-crystal silicon—is lowered until it barely enters the melt. Because the solid seed crystal is at a lower temperature than the melt, heat flows from the melt to the seed, the temperature of the melt in contact with the seed falls, and some of the melt solidifies onto the seed, the atoms arranging themselves to have the same orientation as the atoms in the seed crystal. The seed crystal is rotated (about 60 rpm) and slowly raised (about 1 in. per hr), growing larger as more silicon solidifies onto it. Typical pulled silicon crystals are cylindrical in shape, between 1 and 2 in. in diameter by about 12 in. long.

The impurity dopant to give n-type or p-type silicon is added to the silicon during the initial melting process so that the pulled crystal has the required conduction properties.

In the fabrication of transistors and integrated circuits, the silicon is used in the form of thin circular slices. The single-crystal ingots, grown as above, are sawn into slices about 15 mils thick by a thin, diamond-impregnated wheel rotating at high speed. The slices are then lapped to about 10 mils thickness using a slurry of carborundum powder, and finally the surface of the slice is polished by etching in a concentrated acid mixture. The slices are then ready for the subsequent processes.

Epitaxial Growth.   In the fabrication of solid-state structures, it is often desired to form a thin film of single-crystal silicon with certain conduction properties on the surface of another silicon slice.   The process used is called *epitaxial growth,* and films up to a few tenths of a mil can conveniently be formed.   The starting slice must be single-crystal with the required crystal orientation, and is called the substrate.

Hydrogen gas is bubbled through a volatile silicon compound such as silicon tetrachloride, causing it to vaporize.   The mixture of vapor plus hydrogen is fed to a reaction chamber where the silicon substrate slice is heated to about 1200°C. The silicon tetrachloride dissociates, and silicon is deposited onto the surface of the heated slice to form the epitaxial layer, which grows at about 1 micron (0.04 mil) per min.   The conductivity of the epitaxial layer is controlled and arranged to be either p-type or n-type by introducing the requisite amount of a suitable dopant vapor into the hydrogen stream with the silicon tetrachloride vapor.   The surface of the epitaxial layer conforms closely to the surface of the substrate slice and does not require any additional preparation for subsequent processes.

Solid-state Diffusion.   Solid-state diffusion is a process involving the movement of n-type or p-type impurity atoms into the solid silicon slice.   To achieve this movement, it is necessary to heat the slice to a high temperature, between 800 and 1250°C, in the presence of a controlled density of the impurity atoms, and even then the movement is very slow—a typical rate is 0.1 mil per hr.   Because of the possibility of modifying the conduction properties of very thin layers with very good control, diffusion is a key process in the fabrication of transistors and integrated circuits.

In practice, the process is often carried out in two steps.   The first step consists of heating the silicon slice in the impurity dopant vapor to form a high concentration of dopant on the surface.   This step is called *deposition.*   The slice is then removed to another furnace where it is heated to a higher temperature so that the dopant atoms on the surface move, or diffuse, into the silicon.   This is called the *diffusion* step.

If a p-type impurity is diffused into the surface of an n-type slice such that the density of p-type atoms then exceeds the original density of n-type atoms in the slice, the surface will be changed to p-type, and a p-n junction will be formed a small distance in from the surface where the density of the diffused p-type atoms equals the original n-type density.

It is possible to make a second diffusion into a region that has been formed by a first diffusion.   In this case, it is necessary for the second diffusion to establish an impurity concentration greater than that produced by the first.   Then a second p-n junction will be formed by the second diffusion, and the result of the two diffusions will be an n-p-n structure.

Convenient diffusant impurities for silicon are boron as a p-type impurity and phosphorus as an n-type impurity.

Silicon Oxide Masking.   A most important and significant fact is that a layer of silicon oxide on the surface of a silicon slice will prevent the diffusion of certain elements, including boron and phosphorus, into the silicon.   Also important is that silicon oxide can readily be removed from the surface of the silicon slice by etching with a hydrofluoric acid solution without etching the silicon. Thus, if we oxidize a slice of silicon by heating it in a flow of oxygen to form a layer of silicon dioxide on the surface and then remove the oxide from selected regions by etching, we can arrange to diffuse impurities into these selected regions only. This selective diffusion is the basis of all silicon monolithic integrated-circuit fabrication; it allows the simultaneous formation of a number of separate components in a single slice of silicon.

Oxidation of the silicon slice is carried out by heating at a temperature around 1000°C in a flow of oxygen or steam. The latter is more often used, as it results in a faster rate of oxide growth. Typical oxide thickness is about 10,000 Å (0.04 mil), produced by heating in steam at 1000°C for 4 hr.

Photoresist Process for Oxide Removal. The selective removal of silicon dioxide is carried out by a photolithographic process using photoresist material. The several steps in the process are illustrated in Fig. 2.6. After oxidation (*a*), the oxidized surface of the slice is coated with a thin layer of photoresist lacquer (*b*). This is an organic substance which polymerizes when exposed to ultraviolet light, and then, in that form, it resists attack by acids and solvents. A photographic mask,

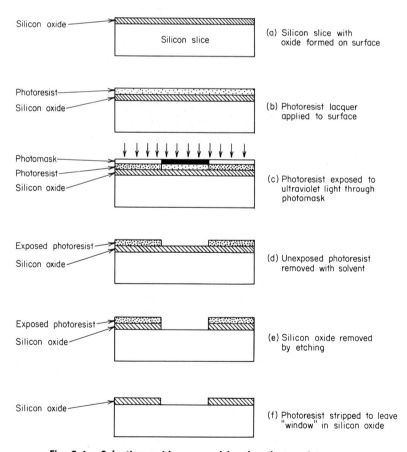

Fig. 2.6.   Selective oxide removal by the photoresist process.

with opaque regions located where it is required to remove the silicon oxide, is placed over the slice and illuminated with ultraviolet light (*c*). The photoresist under the opaque regions of the photomask is unaffected and can be removed with a solvent, the exposed photoresist remaining in the other regions (*d*). The slice is baked to harden the photoresist and then immersed in a hydrofluoric acid solution to etch away the silicon oxide where it is not protected by the polymerized photoresist (*e*). Finally the photoresist is removed from the surface (*f*), and the slice is thoroughly washed. It is now ready for diffusion, which will only occur through the openings (sometimes called *windows*) in the oxide. This complete photoresist process must be repeated each time the silicon oxide is selectively removed.

The Planar Process.   The combination of oxidation, selective oxide removal, and diffusion forms the basis of the *planar process*, which is now firmly established as the basic process of solid-state technology. The sequence of processes used to fabricate a silicon planar n-p-n transistor will be described in some detail. The processes are carried out on whole silicon slices, about 1.5 in. in diameter by 10 mils thick. Each slice normally contains a large number of individual device patterns, and at the end of the slice processing it is cut up into individual wafers.

Referring to Fig. 2.7, an n-type silicon slice is oxidized (*a*), and windows for the base diffusion are opened in the oxide (*b*) by the photoresist process as in Fig. 2.6. Boron is used as the p-type impurity for the base diffusion. Boron tribromide, a liquid, is vaporized, and the vapor is mixed with nitrogen and passed over the silicon slice heated to a temperature of 850°C. During this process, boron is deposited onto the surface of the silicon. The slice is then transferred to another furnace and heated at 1150°C in a flow of nitrogen for a sufficient time (about 1 hr) for the boron to diffuse in and form the p-n junction at the required depth. During the latter part of this diffusion, steam is mixed with the nitrogen so that a new layer of silicon dioxide forms on the surface of the diffused region (*c*). In addition to diffusing down into the silicon, the boron also diffuses sideways, so the p-n junction is formed under the oxide and is protected against surface contamination. This is a very important feature of the planar process. A typical base diffusion depth is 0.1 mil.

The slice is now prepared for the emitter diffusion by etching windows in the new oxide grown over the base region (*d*), using the identical photoresist process as before. To form the n-type emitter region, phosphorus is diffused in. Liquid phosphorus oxychloride is vaporized and passed over the slice at 1000°C. This is usually a single-step diffusion, and for the latter part of the cycle, steam is again introduced to form silicon oxide on the surface (*e*). The emitter diffusion depth is about 0.06 mil, resulting in a base width between the collector and emitter junctions of 0.04 mil.

The next process is to form metallized contacts to the base and emitter regions. Once more the photoresist process is used, and contact windows are opened in the silicon oxide (*f*). Aluminum is now evaporated onto the whole surface of the slice, and a fourth photoresist sequence is carried out with a "reverse" contact photomask to remove the aluminum from everywhere but in the contact windows. The aluminum remaining in the contact windows is then alloyed to the silicon to form a low-resistance contact (*g*).

Finally the slice is cut up into individual transistor elements by scribing between

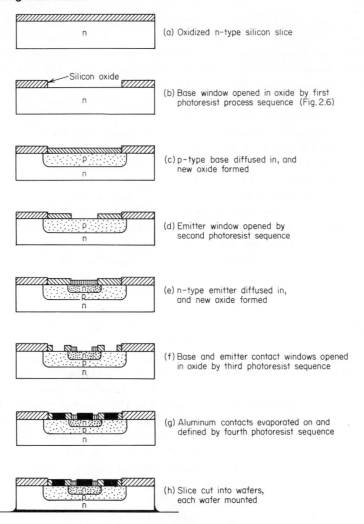

(a) Oxidized n-type silicon slice

(b) Base window opened in oxide by first photoresist process sequence (Fig. 2.6)

(c) p-type base diffused in, and new oxide formed

(d) Emitter window opened by second photoresist sequence

(e) n-type emitter diffused in, and new oxide formed

(f) Base and emitter contact windows opened in oxide by third photoresist sequence

(g) Aluminum contacts evaporated on and defined by fourth photoresist sequence

(h) Slice cut into wafers, each wafer mounted

Fig. 2.7. Steps in the diffused planar process.

the rows of elements and breaking into wafers. The individual wafers are assembled into transistor units by fusing down to a header, which forms the collector contact, and bonding connections to the base and emitter contact areas (*h*).

It will be seen that all of the above processes are carried out on the top surface of the slice, and the three regions of the transistor, the emitter, base, and collector all come to this same plane surface, and hence the name *planar*. By changing the photomask details only, any size and shape can be given to the diffused regions, and so any desired element can be produced with the same basic diffusion processes.

The above description has shown how the planar process is used to fabricate discrete transistors. Diodes can be made by forming the contact areas after stage (*c*). In the next lesson, it will be shown how the planar process is used to fabricate complete monolithic integrated circuits.

THE MOS TRANSISTOR

So far, we have only discussed the *bipolar* junction transistor—bipolar because two types of carrier, the free-electron and the positive-hole, are involved in its operation. A more recently developed transistor, the metal-oxide-semiconductor field-effect transistor (the MOS transistor), is of considerable importance in integrated circuits. The basic operation of this transistor is quite different from that of the bipolar transistor. In it, a conducting channel is induced between two very closely spaced electrode regions by increasing the electric field at the surface of the semiconductor between the electrodes.

The basic structure is shown in Fig. 2.8. The two electrode regions, called the source and drain, are formed by a p-type diffusion into an n-type silicon wafer. Between the source and drain are two p-n junctions back to back, p_1n and np_2. With a voltage V_{DS} applied between the source (positive) and the drain (negative), the np_2 junction is reverse-biased, and so no current will flow from source to drain. If now the gate electrode over the space between source and drain is made sufficiently negative with respect to the source, holes are attracted to the surface of the n-type region and cause it to change to p-type. Then we have two p-type electrodes with a p-type channel joining them, and a current can flow.

The steps in the fabrication of an MOS transistor are shown in Fig. 2.9. An n-type silicon slice is oxidized (*a*), and a photoresist sequence is used to form a window in the oxide for the complete device (*b*). Now a new, thin layer of oxide is formed in the window by oxidation in steam (*c*), and a second photoresist process is used to open windows for the source and drain diffusion (*d*), and boron is diffused in (*e*). The thin oxide is then removed by immersing the slice in a hydrofluoric acid solution (*f*). Next a new, very pure oxide layer is grown over the device region (*g*), and contact windows for the source and drain are opened by another photoresist process (*h*). Finally aluminum is evaporated over the whole slice and removed everywhere but in the source and drain contact windows and in the gate-electrode region by a fourth photoresist process (*i*).

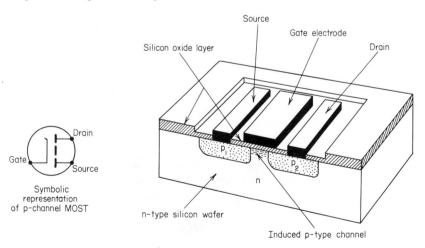

Fig. 2.8. Basic structure of an MOS transistor.

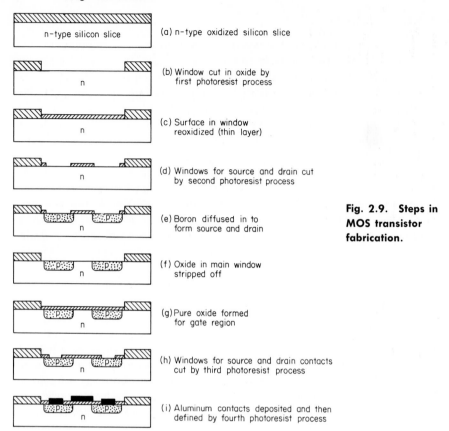

(a) n-type oxidized silicon slice

(b) Window cut in oxide by first photoresist process

(c) Surface in window reoxidized (thin layer)

(d) Windows for source and drain cut by second photoresist process

(e) Boron diffused in to form source and drain

(f) Oxide in main window stripped off

(g) Pure oxide formed for gate region

(h) Windows for source and drain contacts cut by third photoresist process

(i) Aluminum contacts deposited and then defined by fourth photoresist process

Fig. 2.9. Steps in MOS transistor fabrication.

The thickness of the pure oxide under the gate electrode is only of the order of 1000 Å, and the spacing between the source and drain is typically 0.3 mil. The whole structure can be fabricated in an area about 3×1.5 mils, and this makes the MOS transistor very suited for use in integrated circuits where a high density of elements is desired. The fabrication of the MOS transistor in integrated circuits will be discussed in more detail in the next lesson.

GLOSSARY

common-base circuit A transistor circuit in which the transistor is connected so that the base electrode is common to both input and output circuits.

common-emitter circuit A transistor circuit in which the transistor is connected so that the emitter electrode is common to both input and output circuits.

conductor A material in which the outer (valence) electrons are loosely bound to the atom such that they can readily become detached and move under the influence of an electric field to result in a flow of electric current.

crystal pulling A method of preparing a single-crystal ingot of a material in which the developing crystal is slowly withdrawn from a melt.

epitaxial growth   The deposition of a single-crystal film on the surface of a single-crystal substrate such that the crystal orientation of the film is the same as that of the substrate.

insulator   A material in which the outer (valence) electrons are tightly bound to the atom and are not free to move. No current can flow when a voltage is applied across the material.

junction transistor   An active semiconductor device with a base electrode and two or more junction electrodes.

MOS transistor (metal-oxide-semiconductor transistor)   An active semiconductor device in which a conducting channel is induced in the region between two electrodes by a voltage applied to an insulated electrode on the surface of the region.

n-type semiconductor   A semiconductor in which electric conduction is due to the presence of more free electrons than holes.

p-n junction   The region of transition between p-type and n-type semiconductor materials.

p-type semiconductor   A semiconductor in which electric conduction is due to the presence of more holes than free electrons.

planar transistor   A diffused-junction transistor in which the emitter, base, and collector regions all come to the same plane surface, with the junctions between the regions protected at the surface by a layer of material such as silicon oxide.

semiconductor   A material with conductivity roughly midway between that of conductors and insulators and with which the conductivity increases with temperature over a certain temperature range.

single crystal   A piece of material in which all the basic groups of atoms have the same crystallographic orientation.

solid-state device   An electric device which operates by the movement of electrons within a solid piece of semiconductor material.

solid-state diffusion   The introduction of atoms of an impurity element into the surface regions of a solid semiconductor wafer.

REVIEW

For each of the numbered statements below, select the one of the items lettered *a, b, c,* or *d* that correctly completes the statement.

2.1. Monolithic integrated-circuit wafers are typically
   a. 1 in. in diameter.
   b. 0.5 in. square.
   c. $\frac{1}{20}$ in. square.
   d. 5 mils square.

2.2. n-type silicon can be
   a. Pure silicon.
   b. Formed by doping pure silicon with phosphorus.
   c. Formed by doping pure silicon with boron.
   d. Nonconducting.

2.3. When a p-n junction is biased in the forward direction
   a. No current flows.
   b. Only electrons in the n-type side are attracted into the p-type side.
   c. Only holes in the p-type side are attracted into the n-type side.
   d. Both electrons in the n-type side and holes in the p-type side are attracted to the opposite side.

2.4. For normal operation of an n-p-n transistor
  a. The base is negative with respect to the emitter.
  b. The emitter is positive with respect to the collector.
  c. The collector is negative with respect to the base.
  d. The emitter junction is forward-biased, and the collector junction reverse-biased.

2.5. Epitaxial growth is carried out by
  a. Repeated distillation.
  b. Depositing silicon onto a polycrystalline rod.
  c. Depositing silicon from a vapor onto a heated substrate slice.
  d. Pulling a single crystal from the melt.

2.6. Solid-state diffusion
  a. Is a key process in the fabrication of transistors and integrated circuits.
  b. Can be carried out at low temperature.
  c. Cannot be used to form p-n junctions.
  d. Causes impurities to move rapidly into the silicon slice.

2.7. Silicon dioxide is used
  a. As a diffusing element.
  b. As a mask against diffusion of certain elements.
  c. As a contact material.
  d. As a resistor material.

2.8. The photoresist process is used
  a. To photograph the silicon slice.
  b. To prevent photoresponse.
  c. To control the etching of silicon oxide from selected regions on a silicon slice.
  d. During the high-temperature diffusion to prevent diffusion into selected regions.

2.9. The planar transistor
  a. Is a diffused transistor with a layer of silicon oxide on the surface.
  b. Has emitter and collector junctions formed simultaneously.
  c. Is not suitable for integrated circuits.
  d. Has no silicon oxide on the completed structure.

2.10. The MOS transistor
  a. Has only one p-n junction.
  b. Has only two electrodes.
  c. Conducts when sufficient voltage is applied to the gate electrode.
  d. Has a gate electrode in direct contact with the silicon.

Integrated-circuit Technology

INTRODUCTION

The invention of the transistor brought along a considerable revolution in electronics, moving it away from the earlier vacuum-tube techniques. The transistor enabled engineers to produce electronic equipment which was smaller, lighter, more versatile, more reliable, less costly, and required less operating power. But the transistor was only the prelude to a much greater revolution—the monolithic integrated circuit. Integrated circuits performing complete circuit functions in a space comparable with that previously occupied by a single transistor are now becoming the basic components of electronic equipment, taking the place of assemblies of transistors and diodes with resistors and capacitors.

In this lesson it will be shown how the basic technologies described in Lesson 2 are used to fabricate silicon monolithic integrated circuits. The method of forming the various electronic-circuit elements within a wafer of silicon and interconnecting them to give a complete electronic circuit will be described.

MICROELECTRONICS

Since the early days of electronics, there have been continual efforts to miniaturize electronic equipment. The trend was greatly accelerated during World War II, when the need for portable and airborne equipment was emphasized. The advent of the transistor and semiconductor diode after the war encouraged more intensive development of miniature passive components (resistors and capacitors). The use of these miniature components was made possible because the characteristics of transistors allowed circuit operation at very low voltage and power levels, with corresponding low heat dissipation. Assemblies of transistors with the new miniature passive components on small printed circuit boards gave a significant reduction in the size and weight of equipment. The result, though miniature, was still conventional in that an assembly of discrete components was used; it could perhaps be described as *microassembly*. To effect further reduction in size, developments proceeded along basically new lines and led to what is now called microelectronics.

The terms *microelectronics* and *integrated circuits* are sometimes used interchangeably, but this is not strictly correct. Microelectronics is the general name given to extremely small electronic components and circuit assemblies, made by thin-film, thick-film, or semiconductor techniques. An integrated circuit is a special case of microelectronics and refers to a circuit that has been fabricated as an inseparable assembly of electronic elements in a single structure which cannot be divided without destroying its intended electronic function. Thus, integrated circuits come under the general category of microelectronics, but all microelectronic units are not necessarily integrated circuits.

There are two basic approaches to microelectronics, monolithic integrated circuits and film circuits. In *monolithic integrated circuits,* all circuit elements, both active and passive, are simultaneously formed in a single, small wafer of silicon by the diffused planar technique. The elements are interconnected to form the required electronic circuit by metallic stripes deposited onto the oxidized surface of the silicon wafer using evaporation techniques. *Film circuits* are microminiature electronic circuits fabricated by forming the passive electronic components and metallic interconnections directly on the surface of an insulation substrate and subsequently adding the active semiconductor devices, usually in discrete wafer form. There are two types of film circuits, thin-film and thick-film. In *thin-film circuits* the passive components and interconnection wiring are formed on glass or ceramic substrates using evaporation techniques. The active components (transistors and diodes) are fabricated as separate semiconductor wafers and assembled into the circuit. *Thick-film circuits* are prepared in a similar manner except that the passive components and wiring pattern are formed by silk-screen techniques on ceramic substrates.

Other integrated circuits are produced using a combination of techniques. In *multichip circuits,* the electronic components for a circuit are formed in two or more silicon wafers (chips), and the chips are mounted side by side on a common header. Some interconnections are included on each chip, and the circuit is completed by wiring between the chips with small-diameter gold wire. *Hybrid integrated circuits* are combinations of monolithic and film techniques, in which the active components are formed in a wafer of silicon using the planar process, and the passive components and interconnection wiring pattern are formed on the surface of the silicon oxide which covers the wafer, using evaporation techniques.

Although film circuits are being used in significant quantities, the emphasis in this lesson will be on the silicon monolithic integrated circuit, as it is now generally accepted that this will be the preferred approach to microelectronics.

It has become the practice to regard the monolithic integrated circuit as a single electronic component since it is fabricated and installed as a single entity. The circuit components, as they were called in discrete assemblies, are referred to as circuit elements of integrated circuits. From now on we will adhere rigidly to the use of the word element for this purpose.

MONOLITHIC INTEGRATED CIRCUITS

Monolithic integrated-circuit technology is basically an extension of the diffused planar process described in Lesson 2. The active elements (transistors and diodes) and the passive elements (resistors and capacitors) are all formed in the silicon slice by diffusing impurities into selected regions to modify the electrical characteristics and, where necessary, to form p-n junctions. The various elements are designed so that they can all be formed simultaneously by the same sequence of diffusions. In practice, the details of the diffusion processes are decided by the requirements of the transistors, and then the geometry of the other elements is designed so that the desired values are obtained with the transistor diffusion schedules. All process operations are carried out on the top surface of the silicon slice, and all of the element contact regions are formed on this same surface so that they can be interconnected to form the complete electronic circuit by evaporating a metallic wiring pattern on the top of the silicon oxide which covers the surface between the contact areas. As with planar transistors, the processes of selective oxide removal, diffusion, and metallization are carried out on whole silicon slices. On each slice, the same circuit pattern is repeated a large number of times; for example, with a typical integrated circuit having wafer dimensions 50 mils square, a single slice of silicon, 1.5 in. in diameter, will contain about 500 circuits, which are all processed at the same time. The detailed makeup of a single slice of silicon—500 integrated circuits each containing up to 50 circuit elements—is illustrated in Fig. 3.1.

The general sequence of monolithic integrated-circuit fabrication is illustrated graphically in Fig. 3.2. The first step is the "breadboard" design of the electronic circuit using discrete components. The circuit is designed to perform the required function and to ensure that the values of the circuit elements are compatible with the diffusion processes. Next, the circuit elements are designed dimensionally and the complete circuit laid out in a geometric pattern. This is usually carried out by drawing the layout about 500 times full size; that is, a wafer 50 mils square is drawn about 2 ft square. From this drawing, a series of related drawings are prepared, one for each of the oxide-removal steps, and then each of these drawings is reduced to actual size by a series of photographic processes. At the same time as the final reduction to life size, the pattern is repeated by indexing the photographic plate under the image in a *step-and-repeat* sequence, so that for each oxide-removal step, a *master* photographic mask is obtained, containing a matrix of the circuit

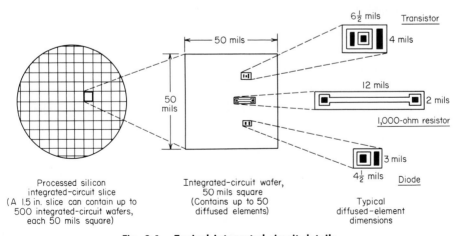

Processed silicon
integrated-circuit slice
(A 1.5 in. slice can contain up to
500 integrated-circuit wafers,
each 50 mils square)

Integrated-circuit wafer,
50 mils square
(Contains up to 50
diffused elements)

Typical
diffused-element
dimensions

Fig. 3.1. Typical integrated-circuit details.

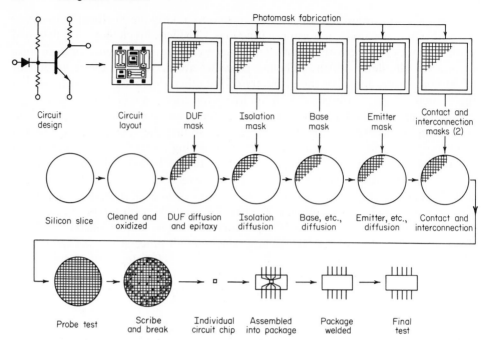

Fig. 3.2. Monolithic integrated-circuit fabrication sequence.

patterns in precise location over an area greater than the slices to be processed. From the master photomasks, copies are prepared by contact printing, and these copies are used to selectively expose the photoresist during the oxide-removal steps.

Starting with a silicon slice about 1.5 in. in diameter and 12 mils thick, the processes of oxidation, selective oxide removal, diffusion into the "windows" so formed, and contact metallization are carried out using the planar process as described in Lesson 2 (Figs. 2.6 and 2.7). The photomasks define the precise location of each element for the oxide removal, and the diffusions modify the properties of the silicon within these locations to form the different elements. A detailed description of the complete process will be given later in this lesson.

INTEGRATED-CIRCUIT-ELEMENT FORMATION

To obtain a good understanding of integrated-circuit fabrication, it is necessary to know how each type of circuit element is formed by the diffusion process. The types of elements used are essentially the same as those used in discrete circuits—transistors, diodes, resistors, and capacitors. A requirement common to all elements in integrated circuits is that, since they are formed in close proximity in a wafer of silicon which is electrically conducting, it is necessary to arrange that they each be electrically isolated from the main part of the silicon wafer so that unwanted coupling between the elements is minimized. Then the only connections between the elements will be the metallized pattern on the surface.

Isolation Techniques.   Several alternative methods of isolation have been developed.   The two most commonly used are diode isolation and oxide isolation.   With *diode isolation* each element is surrounded by a reverse-biased p-n junction.   In this process (Fig. 3.3), an n-type epitaxial layer is first grown on a p-type substrate slice of silicon.   The surface of the epitaxial layer is oxidized (*a*), and the oxide is selectively removed from everywhere but the regions in which the elements will be formed (*b*).   A p-type diffusion is then carried out, and the p-type regions so formed extend down through the epitaxial layer and join up with the p-type substrate (*c*).   This leaves n-type regions, each separated from the substrate by a p-n junction (*d*).   When the final integrated circuit is operated, the p-n junctions are all biased in the reverse direction by connecting the p-type substrate to a potential more negative then any part of the circuit.   Then each junction presents a very high resistance which isolates the element formed in its n-type region.

With *oxide isolation,* a layer of silicon oxide is formed around each element as shown in Fig. 3.4.   A slice of n-type single-crystal silicon is taken, and channels are etched in the surface between the locations planned for each element (*a*).   Then the surface of the slice, including the channels, is oxidized to form a continuous layer of silicon oxide (*b*), and polycrystalline silicon is deposited on top of the oxide in an epitaxial reactor (*c*).   Finally the slice is inverted, and the original silicon is lapped down so that only the regions between the channels are left (*d*).   Each of these is a region of single-crystal silicon isolated by the layer of silicon oxide and supported on the substrate of polycrystalline silicon.

A third system of isolation used for special applications is called *beam-lead isola-*

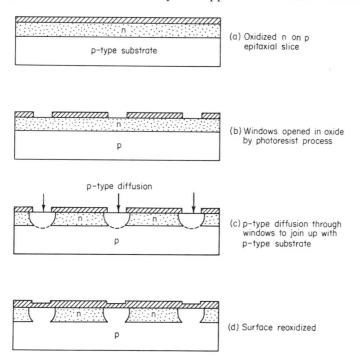

(a) Oxidized n on p epitaxial slice

(b) Windows opened in oxide by photoresist process

(c) p-type diffusion through windows to join up with p-type substrate

(d) Surface reoxidized

Fig. 3.3.   Electrical isolation of integrated-circuit elements by a p-n junction.

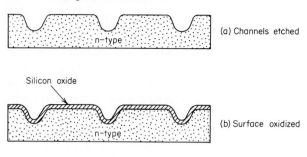

(a) Channels etched

(b) Surface oxidized

Silicon oxide

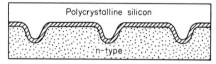

Polycrystalline silicon

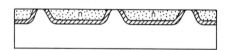

(c) Polycrystalline epitaxial layer deposited on top of oxide

(d) Slice inverted and lapped down

Fig. 3.4. Electrical isolation by a layer of silicon oxide.

tion. The circuit elements are formed in a wafer of silicon in the regular manner. The metallization used for interconnection is increased in thickness over that normally used, and then the silicon between each element is completely removed by etching from the back side. The etchant does not attack the metallization, and the result is that each element is now completely separate, being supported from the top by the metallic connections. A thermosetting plastic can be applied to fill the spaces between the elements to give added mechanical support.

Other methods of isolation are being investigated, and no doubt some will come to fruition, but the three described above will undoubtedly be used for a long time.

Integrated-circuit Transistor Formation. The techniques for fabricating bipolar transistors for integrated circuits are similar to those described in Lesson 2 for discrete planar transistors. A typical arrangement using diode isolation is shown in Fig. 3.5a. After the isolation process, boron is diffused in to form the p-type base region, and then phosphorus is diffused in to form the high concentration n^+-type emitter region. At the same time, another n^+ region is diffused into the n-type collector region so that a low-resistance contact to the collector region can be made. It will be seen that there is one significant difference from the discrete planar transistor in that the collector contact is made at the top surface, alongside the base and emitter contacts. This introduces a problem because the collector current must now flow laterally along the narrow n-type collector region to reach the contact, and so there is additional series-collector resistance compared with the discrete transistor, in which the collector contact is made to the bottom surface. To minimize this series resistance, a low-resistance n^+-type region is selectively diffused into the substrate slice before the epitaxial growth of the n-type layer. This gives the structure shown in

Fig. 3.5*b*. The collector current can now flow straight down into the low-resistance n^+ region and then sideways along it to the vicinity of the contact, resulting in a lower series-collector resistance (R_{CS}). This arrangement is called DUF (diffusion under the epitaxial film).

The equivalent circuit of the transistor including the isolation junction is shown in Fig. 3.5*c*. The isolation junction will have a capacitance C_1 shown in parallel, and the series resistance R_1 is due to the resistance of the substrate between the active transistor region and the substrate contact. At high frequencies and fast switching speeds, the effect of the isolation diode capacitance must be carefully evaluated, as it may be high enough to allow some stray coupling to the substrate and to other elements. With integrated-circuit diagrams it is customary to use the transistor symbol without a surrounding circle as shown in Fig. 3.5*c*, as the transistor has no individual encapsulation.

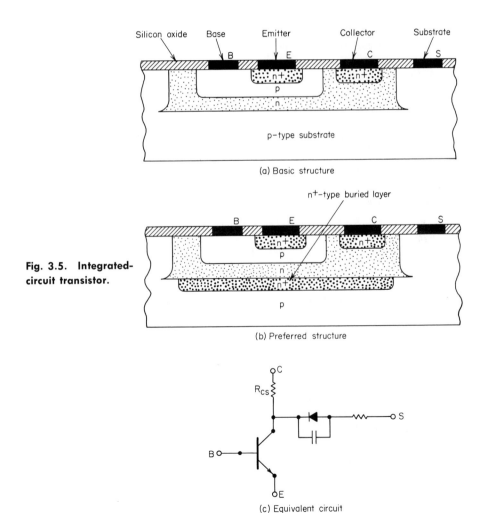

(a) Basic structure

(b) Preferred structure

Fig. 3.5. Integrated-circuit transistor.

(c) Equivalent circuit

MOS Transistors.   The metal-oxide-semiconductor (MOS) transistor as described in Lesson 2 (Fig. 2.8) is fabricated for integrated circuits using the identical processes detailed in Fig. 2.9.   Because of its construction, the MOS transistor is self-isolating. Both the source and drain are isolated by their own p-n junctions, the gate is isolated by the thin layer of silicon oxide, and the channel formed under the gate is also isolated by a p-n junction which forms with it.   This means that MOS transistors can be fabricated in a smaller area than bipolar transistors, allowing a higher element density and a lower manufacturing cost per element.

The MOS transistor can be used as a resistor between source and drain, the value being dependent on the gate potential and the transconductance of the structure. Resistors having values compatible with switching circuits can be obtained by designing the MOS structure to have a low transconductance (wide source-to-drain spacing) and connecting the gate to the drain so that the structure is biased to the ON state.   Such resistors can be made in a much smaller area than that required for diffused resistors, allowing a still further increase in element density.   It should be mentioned that to offset these advantages of high element density and possible low cost per circuit function, the MOS circuit has a considerably slower switching speed than the bipolar circuit.   Further comparison between MOS and bipolar circuits will be made later.

Integrated-circuit Diodes.   Integrated-circuit diodes are prepared by forming p-n junctions at the same time as one of the transistor junctions.   Figure 3.6a shows a diode in which the cathode is the original n-type region, and the p-type anode is formed during the transistor base diffusion.   This type of diode has the same reverse-voltage capability as the transistor collector junction, and is widely used for

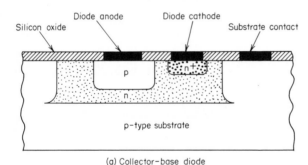

(a) Collector-base diode

Fig. 3.6.   Integrated-circuit diodes.

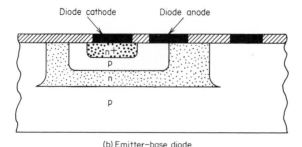

(b) Emitter-base diode

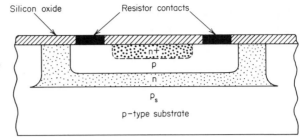

Fig. 3.7. Diffused p-type resistor.

general-purpose circuits. Where fast switching speeds are required, emitter-base diodes are used, as illustrated in Fig. 3.6*b*. The diode anode is formed at the same time as the transistor base, and the cathode with the emitter. This gives a low-voltage diode with fast response time. With this type of diode, to avoid unwanted effects due to transistor action, it is arranged that the anode contact short-circuits the p-type anode region to the n-type region in which the diode is formed.

Integrated-circuit Diffused Resistors. Silicon is a resistive material; its resistivity depends upon the concentration of current carriers (electrons or holes). A resistor can be formed in a silicon wafer by diffusing a suitable impurity into a defined region, the value of the resistor so formed depending on the concentration of the impurity, the dimensions of the region at the surface, and the depth to which the impurity is diffused in. Most resistors in integrated circuits are formed at the same time as the p-type transistor base region. Since the carrier concentration and diffusion depth will be fixed by the requirements of the transistor, the dimensions of width and length of the resistor at the surface must be designed to give the required resistance value. The surface concentration of the transistor p-type base region is typically 100 ohms per square, and with this, a resistor stripe 1 mil wide will have a resistance of 100 ohms per mil of length. Thus a stripe 1 mil wide by 10 mils long will have a resistance of 1,000 ohms. By using several stripes in a series grid form, values up to 20,000 ohms can be made, and short, wider stripes will allow values down to 20 ohms. Figure 3.7 illustrates the cross section of a diffused resistor, formed as above. It will be seen that isolation can be given by the junction $p_s n$ and also by the junction p-n.

When high values of resistance are required, an alternative to increasing the stripe length is to reduce the p-type thickness and effective concentration by diffusing into it an n-type region at the same time as the transistor emitter diffusion, as shown dashed in Fig. 3.7. This method, however, is more difficult to control than increasing the length of the stripe.

For very low values of resistance, the higher-concentration emitter diffusion can be used to form n-type resistors. By this means, values down to 1 or 2 ohms can be obtained.

For various reasons associated with the diffusion process, at the present time it is difficult to reproduce diffused resistors to better than ± 5 percent of the required absolute value. However, the ratio between two resistors formed side by side can be reproduced very accurately, to within ± 1 percent. Thus circuit design for integrated circuits tends to use resistance ratio, rather than the actual resistance value, as a controlling factor. This situation will almost certainly change as improvements in processing gradually allow better reproduceability of diffusion surface concentration.

Integrated-circuit capacitors.   For monolithic integrated circuits, two types of capacitors can be prepared, a junction capacitor or an MOS capacitor.   A *junction capacitor* uses the capacitance of a reverse-biased p-n junction which can be formed at the same time as the emitter junction or the collector junction of the transistor.   The value of capacitance per unit area is quite low, and the maximum value used is limited by economic considerations to the order of 100 pf.   Since the capacitance of a p-n junction depends upon the value of reverse voltage, it will be necessary to arrange the correct voltage bias in the circuit.   Against these limitations, it has the advantage that it can be formed at the same time as the other elements with no additional processes. The arrangement of a capacitor with a junction formed at the same time as the transistor collector junction is shown in Fig. 3.8a.

The structure of an *MOS capacitor* is illustrated in Fig. 3.8b.   An n$^+$ region is diffused into the silicon at the same time as the transistor emitter diffusion to form the bottom electrode of the capacitor, and a controlled thickness of silicon oxide is formed on the surface of this region to give the dielectric.   The top electrode consists of a layer of metal deposited at the same time as the interconnection pattern.   A somewhat higher value of capacitance is possible with this method, but it is still limited to a few hundred picofarads.

COMPLETE INTEGRATED-CIRCUIT FORMATION

In the formation of a complete integrated circuit, the circuit elements are all formed simultaneously by the same sequence of oxidation, selective oxide removal, diffusion,

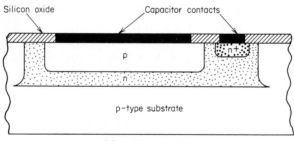

(a) Junction capacitor

Fig. 3.8.   Integrated-circuit capacitors.

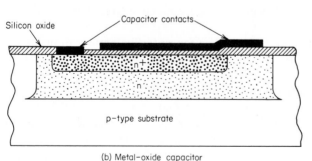

(b) Metal-oxide capacitor

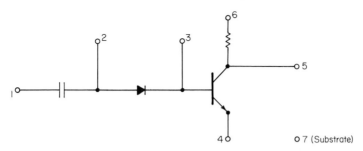

Fig. 3.9. Partial circuit used in the illustration of the fabrication sequence of an integrated circuit shown in Fig. 3.10.

and metallization. The sequence will be illustrated by considering a part of an electronic circuit shown in Fig. 3.9. For convenience it has been assumed that the elements are formed in line; in practice, they may be in any disposition. The several steps in the process are illustrated in Fig. 3.10a.

The process starts with a slice of p-type silicon, which is oxidized on the top surface (a). The first step is the n^+ DUF process to give the low series-collector resistance for the transistor (b). For each of the diffusion steps, the selective oxide removal is carried out using the photoresist process described in Lesson 2 and illustrated in Fig. 2.6. After the n^+ diffusion, the oxide is removed, and an n-type epitaxial layer is grown over the whole surface of the slice (c). It is in this n-type layer that all of the circuit elements will subsequently be formed. The surface is reoxidized, windows are etched in the oxide, and the p-type isolation diffusion is carried out to define the regions of the n-type layer for each element (d). Step 5 is to diffuse the p-type regions for the transistor base, the diode anode, the resistor, and the first capacitor electrode (e). Step 6 is to diffuse n^+ regions for the transistor emitter, the collector contact, the diode-cathode contact, and the second capacitor electrode. Finally, the metallization pattern is deposited and defined to make contact to each of the elements and interconnect them on top of the silicon oxide covering surface of the slice to form the complete circuit. Figure 3.10b shows a plan view of the four steps (d) to (g).

To remind you of the order of dimensions, the overall length taken up by this four-element assembly would normally be about 35 mils, and the width (into the paper) about 6 mils.

At this stage the slice processing is complete. A completed slice is shown in Fig. 3.11. The slice, 1.25 in. in diameter, contains approximately 300 complete integrated circuits. Figure 3.12 shows an enlarged view of one circuit; on this particular slice, each circuit is 60 mils square and contains 8 transistor elements, 12 diode elements, and 12 resistor elements—a total of 32.

COMPARISON BETWEEN BIPOLAR AND MOS INTEGRATED CIRCUITS

It was mentioned earlier that MOS elements occupy a smaller area than bipolar elements. This can be illustrated by considering a simple integrated inverter circuit consisting of a transistor plus its load resistor. In Fig. 3.13a the bipolar arrangement is shown using a bipolar transistor and a diffused resistor. The corresponding MOS

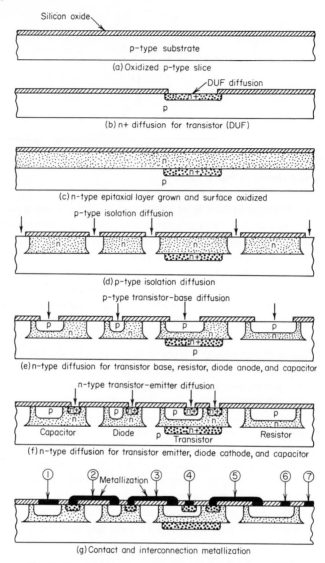

Silicon oxide

p-type substrate

(a) Oxidized p-type slice

DUF diffusion

n+

p

(b) n+ diffusion for transistor (DUF)

n

n+

p

(c) n-type epitaxial layer grown and surface oxidized

p-type isolation diffusion

n   n   n   n

n+

p

(d) p-type isolation diffusion

p-type transistor-base diffusion

p   p   p   p

n   n   n   n

n+

p

(e) n-type diffusion for transistor base, resistor, diode anode, and capacitor

n-type transistor-emitter diffusion

p   n+   p   p   n+ n+   p

n   n   n   n

Capacitor   Diode   Transistor   Resistor

p   n+

p

(f) n-type diffusion for transistor emitter, diode cathode, and capacitor

① ②Metallization ③ ④ ⑤ ⑥ ⑦

n+

(g) Contact and interconnection metallization

Fig. 3.10a.   Integrated-circuit process steps (section).

circuit, using an MOS transistor and an MOS resistor drawn to the same scale, is shown in Fig. 3.13b.   The MOS circuit gives an area saving between 4 and 5 to 1, but as mentioned earlier, this space saving is at the expense of a slower switching speed and a restricted high-frequency performance.

SLICE PROBE TESTING

While still in complete slice form, all of the individual integrated circuits are probe tested.   A typical probe-testing machine has up to 20 pointed probes which can

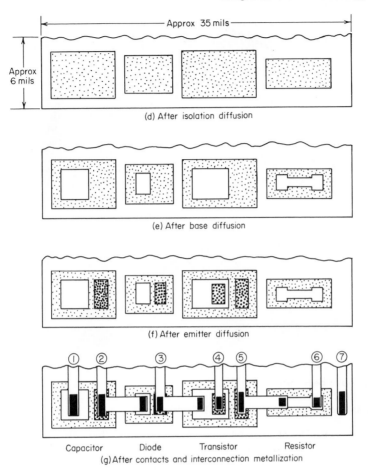

(d) After isolation diffusion

(e) After base diffusion

(f) After emitter diffusion

Capacitor   Diode   Transistor   Resistor
(g) After contacts and interconnection metallization

Fig. 3.10b. Integrated-circuit process steps (plan view).

be positioned individually to an accuracy better than 1 mil and lowered to make electrical contact to the terminal contact pads on the integrated circuit. On a typical circuit wafer 50 mils square, there may be 12 terminal pads, and after the probes have been aligned to the first circuit, they must be raised, the slice stepped one circuit along in sequence, and the probe head lowered each time so that the probes make contact with the pads of each circuit in turn. It will be apparent that probe-testing machines are in the class of high-precision equipment.

The tests carried out at probe are mainly dc tests. A few ac measurements are made, but switching speed and high-frequency tests are limited by capacitance and inductance associated with the construction of the probe head and the connections between the probe head and the measuring circuit. Despite these limitations, it is possible to select good circuits with a probability of about 80 percent. Any circuits failing to meet the test standards are automatically marked with an ink spot so that they can readily be identified and rejected after the slice has been cut up into individual chips.

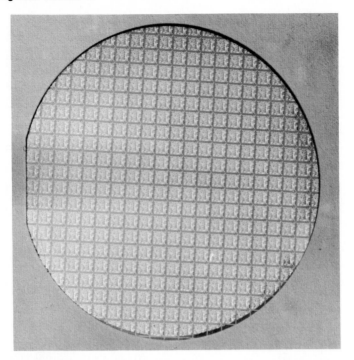

Fig. 3.11. Photograph of a 1.25-in. diameter silicon slice in which 300 integrated circuits have been formed.

SEPARATION INTO INDIVIDUAL CIRCUIT CHIPS

The silicon slice must now be separated into individual integrated-circuit chips. The most common method is *scribing and breaking,* a process very similar to glass cutting. Lines are scribed across the slice between the circuits in each direction using a fine diamond point. Figure 3.14 shows a slice being scribed. Then the slice is placed on a rubber pad, stress is applied by running a roller over it, and the slice breaks up into the individual chips. The chips are sorted by visual inspection to pick out and reject those which were marked with the ink spot during the probe test.

ASSEMBLY PROCESSES

Each integrated-circuit chip is now assembled into a package, sealed, and finally tested. Three main forms of packaging are used: a *multipin TO-5 type* circular package, a hermetically sealed package called a *flat pack* because of its thin rectangular configuration, and a *dual-in-line* plastic molded package. The last two packages are illustrated in Fig. 3.15. In each case, the chip must first be mounted into position in the package. It can be mounted either by soldering down to the base with a suitable metal alloy, or, since electrical contact to the bottom of the chip is not required, a low-melting-point glass frit can be used to "cement" the chip down. Both methods involve heating to a temperature between 300 and 400°C.

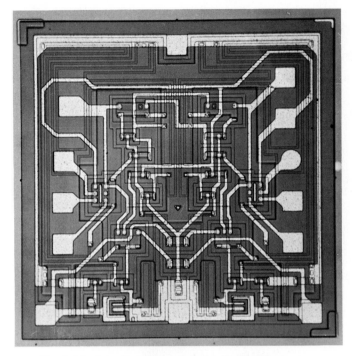

Fig. 3.12. A typical integrated circuit, 60 mils square.

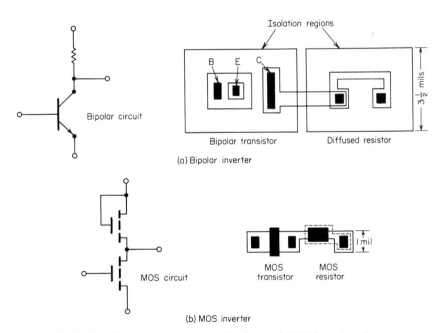

(a) Bipolar inverter

(b) MOS inverter

Fig. 3.13.   Comparison between bipolar and MOS inverter stages.

Fig. 3.14. An integrated-circuit slice being scribed into individual circuits.

    With the chip firmly mounted in the package, the electrical connections from the circuit terminal pads on the chip to the package leads can be made. The most widely used method is thermal-compression bonding. Gold wire about 1 mil in diameter is used with a process called *ball bonding*. The gold wire is fed through a capillary needle as shown in Fig. 3.16*a*, then a minute hydrogen-gas flame is passed across the wire, melting it and forming a ball on the end as shown. The package with the mounted chip is heated to about 300°C, and the capillary is lowered so that the ball on the end of the wire contacts the terminal pad on the chip (*b*). A suitable pressure is applied to flatten the gold ball, and the combination of pressure and temperature results in the gold welding to the circuit pad (*c*). Then the capillary is raised up the wire, moved horizontally until it is over the package terminal, and lowered to weld the wire onto the terminal (*d*). After this second weld, the capillary is raised again, and the wire is "cut" by passing the flame across the wire. This also forms a new ball on the end of the wire, ready to repeat the sequence for the next connection (*e*).

    More recent work is aimed at making all the connections to the chip in a single operation.

    After all the connections have been made in this way, the assembly is ready for finishing. In the case of the flat-pack and TO-5 packages, this consists of welding on a lid to give a hermetic enclosure. With the plastic unit, the assembly is placed in a mold, and the plastic body is molded around it. It is usual to carry out a "leak" test on the devices after sealing to check that they are completely airtight.

The last step in the manufacture is the final test, in which a series of electrical measurements are carried out to determine whether the performance of the circuit is up to the required standard. The nature of the final test will depend upon the type of circuit, but it will be a combination of dc and ac measurements and functional performance of the complete circuit.

YIELD CONSIDERATIONS

It will have been seen that there are many sequential steps in the process of fabricating an integrated circuit. In most of these steps, there is some loss. Such yield losses occur at each of the oxide-removal and diffusion steps due to a variety of causes, such as imperfections in the original silicon, incomplete cleaning of the

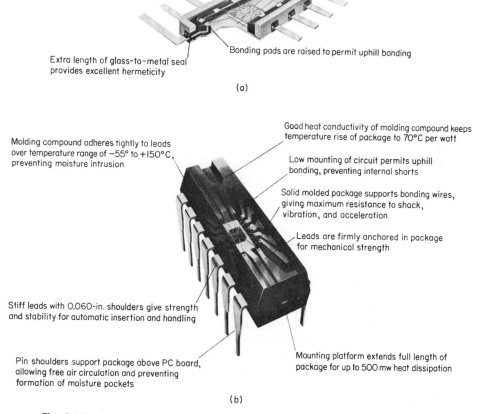

Fig. 3.15. Integrated-circuit package details: (a) flat pack; (b) dual in line.

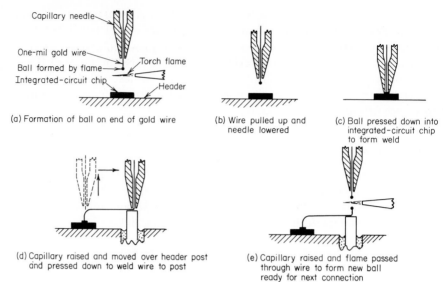

Fig. 3.16. Sequence of the ball-bonding process.

slices, uneven photoresist application and removal, the presence of dust particles and unwanted impurities contaminating the diffused areas, incomplete control over the etching processes, and mechanical breakage of the slices. Although the loss at each operation is small, only 1 or 2 percent, there are so many sequential operations that the cumulative good yield up to the end of scribing the slice into chips and sorting can be quite low—between 10 and 40 percent depending on the circuit. After this, units may be damaged during assembly, and there will be a further loss at the final test due to units not meeting specification; the overall yield can be as low as 5 percent or perhaps up to 20 percent, depending on the type of circuit.

This yield would appear to be low compared with discrete-component electronic assembly, but since on one slice up to 500 circuits are fabricated simultaneously, the economics are such that the final cost is substantially lower than corresponding discrete component assemblies.

The fact that yields are so low means that there is considerable scope for yield improvement and cost reduction in the future.

THIN- AND THICK-FILM CIRCUITS

To complete the picture of microelectronics, a short description of film circuits will be given. To some extent, one can consider both thin- and thick-film circuits as extensions of the earlier discrete-component electronics in that they are assemblies of components either formed on or fixed onto the surface of an insulating substrate. Both methods use techniques that have long been established in other applications, particularly evaporation and silk-screen deposition. However, it could be said that the impetus for the development of thin- and thick-film circuits came from the early success of solid-state techniques, such as the availability of microminiature encapsulated transistors, followed by planar passivated transistor wafers. Also, the general impact of the monolithic integrated circuit simulated alternative developments by electronic-component manufacturers not established in semiconductor techniques.

Thin-film circuits are prepared by evaporation or sputtering processes and use thickness between 0.001 and 0.1 mil. Substrates are usually of ceramic, and size ranges from about 0.25 to 1 in. square. Resistors are made by evaporating nichrome or tantalum in the form of stripes between two high-conductance terminal regions. By varying the length, width, and film thickness of the stripes, both materials can be used to fabricate resistors in the range of 10 ohms to 1 megohm. Thin-film capacitors take the form of two conducting areas separated by a thin layer of insulating dielectric. The required value of capacitance is obtained by choice of dielectric material, its thickness and the electrode dimensions. Typical dielectric materials are tantalum oxide, aluminum oxide, and silicon oxide. Some manufacturers use the material tantalum for all resistor and capacitor formation. A tantalum film is sputtered all over the substrate; the pattern of resistors and capacitors is photoetched into the sputtered film which is then oxidized to form tantalum oxide for the capacitor dielectric. The top electrode for the capacitors and the conducting pattern to interconnect the components is then deposited, using gold or platinum. Discrete active semiconductor components were originally completely encapsulated, miniature transistors and diodes soldered into the circuit, but it is now more usual to use silicon diffused planar transistor and diode wafers mounted onto suitable conducting pads and connected into the circuit by thermo-compression bonding. A typical thin-film circuit is shown in Fig. 3.17.

Thick-film techniques normally relate to methods using the silk-screen process for forming conducting lead patterns and passive components. A typical substrate is alumina, about 0.5 in. square and 60 mils thick. The fabrication process starts with

Fig. 3.17. A typical thin-film circuit.

, LIBRARY OF
JOHN D. ANDERSON

Fig. 3.18. A typical thick-film circuit module: (a) general view, substrates approximately 0.5 in. square; (b) undersurface, showing resistors; (c) top surface, showing transistors and diodes (black) and capacitors (white).

the screening on of a metallized ink interconnection pattern, which is then fired at a temperature of about 700°C. For the resistors, certain types of metal-glass slurry are used. The slurry is applied through a silk screen to form the resistor pattern, and the unit is again fired at about 700°C. The resistor values are subsequently trimmed to 1 percent accuracy by airbrasive removal. Capacitors are either miniature discrete components soldered into the circuit for high values or film-type fabricated on the substrate for low values. For the latter, a platinum bottom electrode is first deposited, and then a dielectric consisting of a mixture of glass and ceramic is applied in the form of a paste and fired. Platinum is again used for the top electrode. The interconnection pattern is coated with solder to a thickness of 2 to 3 mils.

Active devices are in the form of single-sided planar transistor and diode wafers, which are inverted and fused directly to the lead patterns by means of special contact arrangements. Sometimes, the semiconductor wafer is first mounted onto a small

ceramic mount which has metallized projecting areas which register with the wiring pattern on the substrate. A typical thick-film circuit is shown in Fig. 3.18.

GLOSSARY

active device A device displaying gain or control, such as a transistor or vacuum tube. Diodes are generally included in this category, as they can be used to control the flow of current.

beam-lead isolation The method of producing electrical isolation between integrated-circuit elements by interconnecting the elements with thick gold leads and selectively etching away the silicon between elements without affecting the gold leads. This leaves the elements as separate units supported by the gold leads.

diode isolation The method of producing a high electrical resistance between an integrated-circuit element and the substrate by surrounding the element with a reverse-biased p-n junction.

DUF (diffusion under the expitaxial film) A method of introducing a low-resistance path between the active region of an integrated-circuit transistor and the collector contact electrode at the surface. A high-conductance region is formed by selective diffusion in the required location before the epitaxial layer is deposited.

hybrid integrated circuit A combination of different types of integrated electronic parts to give a complete circuit.

junction capacitor A capacitor utilizing the capacitance of a reverse-biased p-n junction.

microelectronics The name given to all techniques used to fabricate very small electronic circuits and systems. It mainly covers all types of silicon integrated circuits, thin-film circuits, and thick-film circuits.

monolithic integrated circuit An electronic circuit which has been fabricated as an inseparable assembly of circuit elements in a single structure which cannot be divided without permanently destroying its intended electronic function.

MOS capacitor A capacitor formed by depositing a silicon oxide dielectric layer and then a metal top electrode on the surface of a conducting semiconductor region which forms the bottom electrode.

multichip integrated circuit An electronic circuit consisting of two or more semiconductor wafers containing single elements or simple circuits, interconnected to give a more complex circuit and encapsulated in a single package.

oxide isolation The method of producing electrical isolation of a circuit element by forming a layer of silicon oxide between it and the substrate.

passive device A device not displaying gain or control, such as a resistor or capacitor.

substrate The physical material on which an integrated circuit is fabricated. Its primary function is for mechanical support, but it may serve some electrical function also.

thick-film circuit A small electronic circuit assembly in which the passive circuit elements and the interconnections between elements are defined on a ceramic substrate using the silk-screen process, and subsequently fired. The active elements are added as discrete chips.

thin-film circuit A small electronic circuit assembly in which the passive circuit elements and the interconnections are formed by evaporation onto glass or ceramic substrates, and the active devices are assembled on as discrete chips.

REVIEW

For each of the numbered statements below, select the one of the items lettered *a, b, c,* or *d* that correctly completes the statement.

3.1. Microelectronics refers to
  a. Circuits using only miniature discrete components.
  b. Monolithic integrated circuits only.
  c. Very small electronic circuits made by thin-film, thick-film, or semiconductor techniques.
  d. Electronic circuits using subminiature tubes.

3.2. In a monolithic integrated circuit
  a. All circuit elements are formed in a single wafer of semiconductor material.
  b. Each circuit element is fabricated as a separate wafer.
  c. Active circuit elements are assembled as separate wafers.
  d. The circuit elements are assembled onto ceramic substrates.

3.3. A multichip circuit
  a. Consists of several interconnected thin-film circuits.
  b. Is a monolithic silicon wafer with thin-film components.
  c. Consists of several interconnected monolithic circuit wafers.
  d. Is normally made with each circuit element on a separate chip.

3.4. In a silicon monolithic integrated circuit, isolation
  a. Is not necessary because the silicon substrate is an insulator.
  b. Is necessary because the silicon is electrically conducting.
  c. Is obtained by mounting separate chips on an insulating base.
  d. Can be obtained by scribing.

3.5. During the slice processing of an integrated circuit
  a. Circuit elements are formed in areas where the silicon oxide is left on the surface.
  b. Only one diffusion process is used.
  c. The diffusing elements diffuse through the silicon oxide.
  d. Circuit elements are formed in areas where the silicon oxide has been removed.

3.6. Photomasking
  a. Controls the depth of diffusion.
  b. Is used to prevent ambient light shining on the silicon slice.
  c. Is used in the process to remove selected regions of silicon oxide.
  d. Reduces the size of the circuit elements.

3.7. Transistors for monolithic integrated circuits
  a. Are identical with discrete planar transistors.
  b. Use the isolation junction as the collector junction.
  c. Are made as separate wafers.
  d. Are similar to discrete planar transistors, but have the collector contact on the top surface.

3.8. In monolithic integrated circuits, diodes
  a. Are double-ended.
  b. Are formed at the same time as either the collector or emitter junction of the transistor.
  c. Are formed on top of the silicon oxide surface of the wafer.
  d. Need a special diffusion process.

3.9. Diffused resistors for integrated circuits
  a. Can be accurately reproduced in absolute value.
  b. Do not need isolation.

    c. Can be produced with an accurate ratio between values.

    d. Can be made with values above 1 megohm.

3.10. Capacitors for integrated circuits

    a. Can be made using silicon oxide as the dielectric.

    b. Cannot be made using diffusion techniques.

    c. Can be made with very high values of capacitance.

    d. Are not possible.

Digital Logic Circuits

INTRODUCTION

Integrated circuits have found their greatest use so far in digital applications, especially in digital computers where there is a requirement for large numbers of identical logic circuits. In 1967, 80 percent of all integrated circuits used were of the digital type. The digital application came with the development of logic circuits for computers, in which a series of simple "yes-no" types of decisions are made by electronic circuits.

The first electronic computers used the decimal system, but this proved to be very cumbersome, since ten distinct levels are required for each order. The problem of defining and maintaining these ten levels proved to be so great that the decimal system was abandoned and a simple binary system in which it is necessary to define only two levels was adopted. The binary system uses only two digits, 0 and 1. How this system is used will be described in the subsequent sections of this lesson. The important point is that in binary arithmetic only two levels are used; for example, a quantity either exists, or it does not exist. This type of decision making is relatively easy to carry out with simple basic transistor circuits, which can make a voltage exist at the output or not. Also, the transistor can change from one condition to the other in less than one-millionth of a second, and so it can make millions of decisions per second.

The basic operation carried out in the logic circuits of a digital computer is the process of arithmetic addition; the processes of subtraction, multiplication, and division are carried out by modifications of the addition process. For example, to multiply two numbers together, the computer adds the first number to itself the number of times equal to the second number. To multiply 15 by 5, the digital computer adds $15 + 15 + 15 + 15 + 15$. It will be apparent that for this type of operation many of the binary circuits will be required, but the simplicity of the system, coupled with the favorable economics of the transistor and the integrated circuit, have made comprehensive computer systems quite attractive commercially. Since this type of computer uses only two digits, it is, in fact, called a *digital computer*. Data are applied in the form of electric pulses of the two discrete voltage levels; the information is represented by the number of pulses and the time spacing between them. In most computers, the data are fed into the computer in decimal form, then are converted to the equivalent binary form, the arithmetic is carried out with binary numbers, and the result is converted back to decimal form for the output.

There is another type of computer called an *analog computer*. Here, the input data are in the form of varying voltages which represent the required quantities by analogy; for example, a varying temperature can be represented by the voltage output from a thermocouple, and the angular displacement of a shaft by a voltage from the slider of a potentiometer. The input data are then manipulated in the computer by linear voltage amplifiers, voltage dividers, etc., to give the required output.

In this lesson, we will be concerned with digital operation, and we will consider the binary system, logic functions based on binary operation, and basic logic circuits as used in digital integrated circuits for application in computers and the like.

THE BINARY SYSTEM

In order to understand the operation of digital integrated circuits, it is necessary to become familiar with the binary system and how it is used in logic decision making. The simplicity of the binary system is illustrated in our everyday life. A switch is open or closed, a light bulb is on or off, a person is present or absent. It is very easy to decide which of the two states exists. For convenience, in computer language the first (or OFF) state is called 0, and the second (or ON) state is called 1, and these are the only two numbers used in the arithmetic unit of a digital computer. At first sight it might seem very difficult and lengthy to carry out the familiar calculations of addition and multiplication with only two numbers, and, in fact, for us to try to use this system for our manual paper arithmetic would be a problem. But you will see how binary digital circuits can be arranged quite conveniently and economically for electronic computer systems.

In our familiar decimal system, we have ten numbers—0 through 9. We first count units up to 9. Then, for the next order, we have to go back to the start with unit 0, but we insert a 1 in the second-order column to indicate that we have counted through all the units once. This gives us 10. Then we count through the units again, indicating the count in the units column until we reach 19. On the next count, the unit must go back to 0 again, and we change the 1 to 2, signifying that we have counted through the units twice, giving us 20. This process is repeated until we reach 99, when we must go back to 0 in both columns, and we indicate this by putting a 1 in the third-order column, giving 100, and so on.

To count with the binary scale we follow exactly the same procedure, using only the numbers 0 and 1. We start with zero and indicate this by 0, and the next count is 1. But now we have used all our numbers, and so for the next count we must go back to 0 and put a 1 in front of it in the second-order column to indicate that we have counted through our scale once. Thus, the number 2 in the decimal system is indicated by 10 (call it one-nought, not ten) in the binary scale. The next count will be indicated by changing the 0 to 1, and we have 11 (one-one), corresponding to 3. Now we have used all our numbers again, and for the next count both columns

Table 4.1.   Equivalent Decimal and Binary Numbers

Decimal number	Equivalent binary number	Power
0	0	
1	1	
2	10	2^1
3	11	
4	100	2^2
5	101	
6	110	
7	111	
8	1000	2^3
9	1001	
10	1010	
11	1011	
12	1100	
13	1101	
14	1110	
15	1111	
16	10000	2^4
32	100000	2^5
64	1000000	2^6
128	10000000	2^7

must go back to 0, and we put a 1 in the third-order column giving 100 (one-nought-nought) as the binary equivalent of 4.   Table 4.1 gives the equivalent binary numbers for decimal numbers up to 16.   The reader should carry on and see how the binary count continues.   Always proceed by adding 1 to the right-hand unit column and carrying 1 to the left as required.   Thus, any decimal number can be expressed in its equivalent binary form and vice versa.

If we examine Table 4.1, we see that the binary number 10 is equal to decimal 2, which is 2^1; the binary number 100 is equal to decimal 4, which is 2^2; the binary number 1000 is equal to decimal 8, which is 2^3; the binary number 10000 is equal to decimal 16, which is 2^4.   Every additional order of the binary number corresponds to an additional power of 2.   This fact is used in converting a binary number to its decimal equivalent.   Take the binary number 11010.   This is equivalent to $2^4 + 2^3 + 0 + 2^1 + 0$ or $16 + 8 + 0 + 2 + 0$ which equals 26.   Conversely, a decimal number can be converted to binary by repeatedly subtracting the highest possible power of 2.   Take the decimal number 26.   We can first subtract 16 (2^4), which in binary is 10000.   From the remaining 10 we can subtract 8 (2^3), which in binary is 1000.   This leaves 2, which in binary is 10.   Adding the binary numbers we have $10000 + 1000 + 10$, or 11010.

The binary numbers require a longer sequence of digits than the decimal equivalents, especially for high numbers.   For example, the binary equivalent of decimal 1,048,576 (2^{20}) is 100,000,000,000,000,000,000.   However, because the electronic digital computer can process millions of simple additions per second, the relative complexity of binary numbers is not difficult to handle.

To carry out the process of addition with binary numbers, we only have to recognize and manipulate the two digits. How this is achieved will be discussed in the following paragraphs.

BINARY LOGIC

Logic circuits are used to make the series of decisions necessary to obtain the logical answer to a problem with a given set of conditions. To make logic decisions, three basic logic circuits have been established, the OR circuit, the AND circuit, and the NOT circuit. These will be discussed in turn.

The OR Circuit. The OR circuit has two or more inputs and a single output. The inputs and the output can each have one of two states, 0 or 1. The circuit is arranged so that the output is in state 1 when any one of the inputs is in state 1; that is, the output is 1 when input A OR input B OR input C is 1. The circuit can be illustrated by the analogy shown in Fig. 4.1a. This is a simple electric circuit consisting of a battery supplying a lamp L through three switches in parallel. We can think of the switches being three inputs to the lamp and the light from the lamp as the output. It will be seen that the lamp will light if switch A OR switch B OR switch C is closed. If we define an open switch and no light as 0 states and a closed switch and a glowing lamp as 1 states, we can list the various combinations of switch states and the resulting output state. This is called a *truth table* and is shown in Fig. 4.1b. It is a convenient way of indicating the overall operation of the circuit. In addition to the ON definition of an OR circuit stated above, it follows that, for there to be no light (output 0), *all* switches must be open (in 0 state).

This type of circuit is called an *OR gate,* and has the symbolic representation shown in Fig. 4.1c, which represents an OR gate with three inputs.

Thus the OR gate is used to make a logical decision about whether at least one of several inputs is in the 1 state.

The AND Circuit. The AND circuit similarly has several inputs and only one output, but in this case the circuit is arranged so that the output is in state 1 only if *all* inputs are in state 1 simultaneously. This is illustrated in Fig. 4.2a. Here it will be seen that the lamp L will light only if switch A AND switch B AND switch

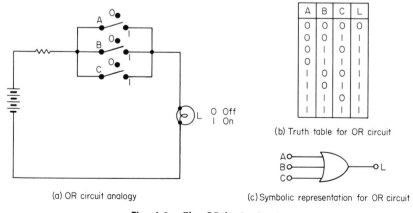

A	B	C	L
0	0	0	0
0	0	1	1
0	1	0	1
0	1	1	1
1	0	0	1
1	0	1	1
1	1	0	1
1	1	1	1

(b) Truth table for OR circuit

(a) OR circuit analogy

0 Off
1 On

(c) Symbolic representation for OR circuit

Fig. 4.1. The OR logic circuit.

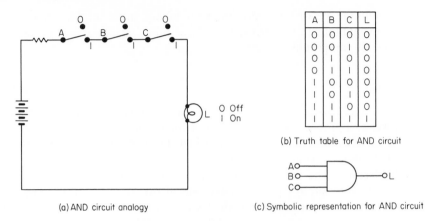

(b) Truth table for AND circuit

(a) AND circuit analogy

(c) Symbolic representation for AND circuit

Fig. 4.2.   The AND logic circuit.

C are all closed at the same time.   The lamp will not light if any one of the switches is not closed.   Using the same notation as before, the truth table for the AND circuit is shown in Fig. 4.2b.   The symbolic representation is shown in Fig. 4.2c.

It will be seen that the AND gate can be used to make a logical decision about whether several inputs are all in the 1 state at the same time.

The number of inputs to a gate is called the *fan-in*.   In the above examples, each gate has a fan-in of three.   There is only one output signal from a gate, but it may be required that this signal be fed to several other gates for further decision making.   The number of subsequent gates that the output of a particular gate can "drive" is called the *fan-out*.

The NOT Circuit.   The NOT circuit has a single input and a single output and is arranged so that the output is always in the opposite state to that of the input.   Consider Fig. 4.3a.   When the switch is open (state 0), current will flow through the lamp, and it will light (state 1).   If the switch is closed (state 1), current will now flow through the switch rather than the lamp, which will go out (state 0).   This operation of making the output state opposite to that of the input is called *inversion*, and a circuit designed to do this is called an *inverter*.   The simple truth table for an inverter is shown in Fig. 4.3b, and the symbol for the circuit in Fig. 4.3c.   The significance and use of the NOT or inverter circuit will become more apparent as we proceed through the subsequent sections of the lesson.

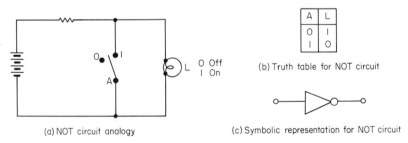

(b) Truth table for NOT circuit

(a) NOT circuit analogy

(c) Symbolic representation for NOT circuit

Fig. 4.3.   The NOT logic circuit.

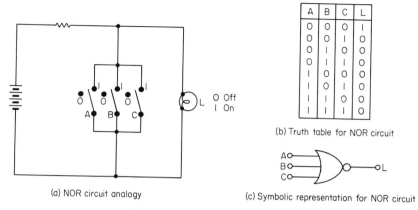

A	B	C	L
0	0	0	1
0	0	1	0
0	1	0	0
0	1	1	0
1	0	0	0
1	0	1	0
1	1	0	0
1	1	1	0

(b) Truth table for NOR circuit

(a) NOR circuit analogy   (c) Symbolic representation for NOR circuit

Fig. 4.4.   The NOR logic circuit.

Combined Logic Circuits.   A NOT circuit can be combined with an OR gate or an AND gate so that inversion takes place along with the other function of the gate. A NOT circuit combined with an OR gate is called a NOR gate (*Not OR*). This is illustrated using our lamp-circuit analogy as shown in Fig. 4.4a. If any one of the switches is in the 1 state, the lamp will be in the 0 state. The truth table is shown in Fig. 4.4b, and the symbol in Fig. 4.4c.

Similarly, a NOT circuit combined with an AND gate is called a NAND gate (*Not AND*), and is illustrated in Fig. 4.5a. When *all* switches are in the 1 position, the lamp will be in the 0 state. The truth table for the NAND circuit is shown in Fig. 4.5b, and the symbol in Fig. 4.5c.

A system of algebra in which the functions performed by binary circuits and systems can be expressed in a simple form to facilitate the analysis of complex logic systems has been established and is called *Boolean algebra* (after the inventor George Boole). This system has been treated in many textbooks and will not be introduced in this course.

Table 4.2 summarizes the input and output states for the various gates.

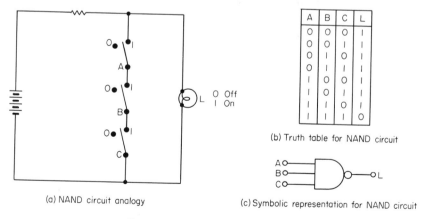

A	B	C	L
0	0	0	1
0	0	1	1
0	1	0	1
0	1	1	1
1	0	0	1
1	0	1	1
1	1	0	1
1	1	1	0

(b) Truth table for NAND circuit

(a) NAND circuit analogy   (c) Symbolic representation for NAND circuit

Fig. 4.5.   The NAND logic circuit.

Table 4.2. Input and Output States for the Basic Gates

Gate	Inputs	Output
AND	All inputs 1	1
	Any one input 0	0
NAND	All inputs 1	0
	Any one input 0	1
OR	Any one input 1	1
	All inputs 0	0
NOR	Any one input 1	0
	All inputs 0	1
NOT (Inverter) . .	Input 1	0
	Input 0	1

USE OF LOGIC GATES IN BINARY ADDITION

To illustrate the use of logic gates, we will consider the operation of adding together two binary numbers A and B. First, we will consider the simplest case, when A and B each consist of one binary digit—either 0 or 1. The logic diagram of a circuit that will carry out the addition is shown in Fig. 4.6a, and the overall truth table is given in Fig. 4.6b. The two inputs to the circuit, A and B, are each connected to an AND gate, A to AND_1, B to AND_2; they are also each connected to the opposite gate through inverters I_1 and I_2. Thus, when A is 1, its input to gate AND_1 is 1, and its input to gate AND_2 is 0; when A is 0, its input to AND_1 is 0, and to AND_2 1. The outputs from the two AND gates are connected to an OR gate, and the output from the OR gate gives the sum S. A and B are also both fed as direct inputs into a third AND gate AND_3, and the output from this gives the carry C.

Consider the operation of the circuit. There are four possible situations.

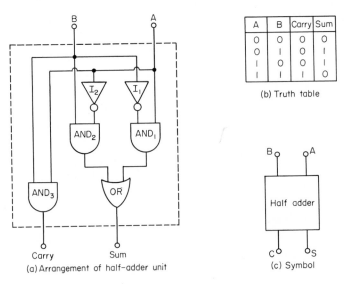

(b) Truth table

(a) Arrangement of half-adder unit

(c) Symbol

Fig. 4.6. A half-adder unit.

1. $A = 0$ and $B = 0$. None of the AND gates can give an output, since at least one input to each of them will be a 0. Thus, both the sum and carry will indicate 0, giving the answer 00.

2. $A = 1$ and $B = 0$. A gives a 1 into gate AND_1, and the 0 at B is inverted by I_1 to give another 1 into gate AND_1. Thus both inputs to AND_1 are 1, and the gate will operate to give a 1 output into the OR gate. Both inputs to gate AND_2 will be 0, and so the output from this gate will be 0. As one of the inputs to the OR gate is 1, the output will give a 1 as the sum. The carry gate AND_3 will not operate as one of its inputs will be 0, and so the carry output will be 0 giving the answer 01.

3. $A = 0$ and $B = 1$. The operation is the same as for situation 2, with the inputs to AND_1 and AND_2 reversed. Again the answer is 01.

4. $A = 1$ and $B = 1$. Neither gate AND_1 or AND_2 can give an output, since one of their inputs will be a 0 from the inverter, and so the sum will be 0. Both inputs to the carry gate AND_3 will be 1, and so it will operate to give a carry output of 1. Thus the answer is 10.

This circuit can be considered as a basic logic-block unit as shown by the dashed lines in Fig. 4.6a, with two inputs and two outputs. As such it is called a *half adder*—*half* because it only adds the two numbers, and when adding two numbers in an order higher than the first, it is necessary to make provision for the circuit to accept and add in a carry from the previous order. To do this a *full-adder* circuit is necessary. One method of making a *full adder* is to use two half adders as shown in Fig. 4.7a. The first adds A and B, and the second adds the resulting sum to a carry input from the previous order to give the final sum. The carry outputs from the two half adders are fed to an OR gate, and the output from this gives the final carry. The truth table for the full adder is given in Fig. 4.7b. It can be seen by

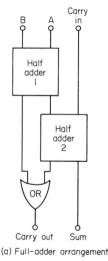

(a) Full-adder arrangement

Carry in	B	A	Carry out	Sum
0	0	0	0	0
0	0	1	0	1
0	1	0	0	1
0	1	1	1	0
1	0	0	0	1
1	0	1	1	0
1	1	0	1	0
1	1	1	1	1

(b) Truth table

Fig. 4.7. A full-adder system.

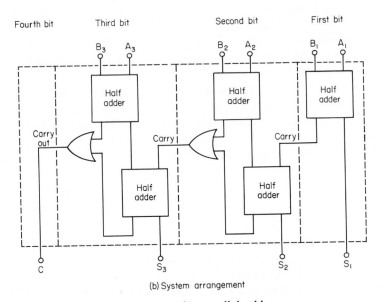

(a) Addition table

(b) System arrangement

Fig. 4.8. A 3-bit parallel adder.

analysis that a carry output cannot exist at the outputs of both half adders simultaneously.

By arranging a series of full adders in parallel, the system can be extended to add binary numbers of several orders. The first order will not require any carry input and so can be a half adder. Figure 4.8b shows the block diagram for a system to add together two 3-order binary numbers, $A_3A_2A_1$ and $B_3B_2B_1$, as indicated in the addition table Fig. 4.8a.

It is interesting to count the number of gates required for these adding units. For a single full adder as shown in Fig. 4.7, we need six AND gates, three OR gates and four inverters—a total of 13 gates just to add two binary digits in the same order. Now modern computers are arranged to handle about 10 decimal orders—decimal numbers up to 10,000,000,000 (2^{33}). Thus, in binary working, we may need 33 orders or bits. To add together two 33-bit binary numbers, we will need 32 full adders and a half adder—a total of 422 gates. With the requirement of repeated addition for multiplication, together with facilities for other manipulations, it is easy to see why the number of gates in the arithmetic unit of a modern digital computer can be up to 10,000. In such an arithmetic unit, there will be a repeated use of the same type of gate; all the AND gates can be identical, and likewise all the OR gates and inverters. It is this that makes integrated circuits so important for digital applications. With integrated circuits, many identical circuits are manufactured simultaneously on one slice of silicon, resulting in identical performance between circuits and low manufacturing cost. Also, the design and tooling costs for a new type of gate are spread over so many manufactured units that they are soon recovered.

In the above description of binary addition, we only use "boxes" labelled AND, OR, etc.; the overall function is obtained by the manner in which these boxes are interconnected. This is an important aspect of logic-circuit design. If one accepts that the gates will operate satisfactorily on a 0 to 1 basis, the design of the logic system can be completely carried out on paper. In fact, it does not matter from the logic-system viewpoint what the boxes contain—relays, vacuum tubes, magnetic cores, transistors, or integrated circuits—the overall logic function will be the same. Which of these devices are used is determined by other things, such as cost, size, power requirement, speed, and reliability. It is in these aspects that the integrated circuit has so much to offer for digital operation.

All digital integrated circuits are based on the use of the transistor as a binary device, and so this will now be discussed in some detail.

THE TRANSISTOR AS A BINARY DEVICE

To carry out the logic functions described above, we need a device that can change between two stable states of electrical conduction. An ordinary light switch does this, but it would be of little use in logic circuits, since it requires mechanical operation, is relatively slow in changing states, and the contact resistance tends to be unstable when the switch is passing small electric currents. The transistor can be used as an electronic switch without any of these limitations. Consider the basic common-emitter transistor circuit shown in Fig. 4.9a. The collector characteristics of the transistor are shown in Fig. 4.9b, and drawn in is the load line which starts at the supply voltage V_{CC} and goes with an inverse slope equal to the value of the load resistor R_L. The intercept of the load line on the current axis is that current which would flow through the load resistor if the transistor were shorted out. If V_B is 0, there will be no base current flowing into the transistor, and so no collector

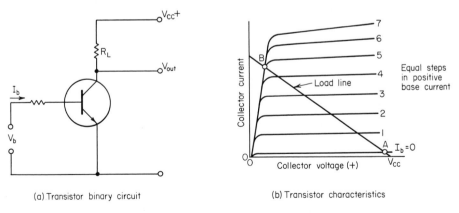

(a) Transistor binary circuit   (b) Transistor characteristics

Fig. 4.9. The transistor as a binary switching device.

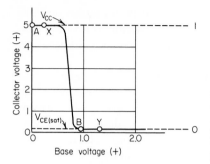

Fig. 4.10. Voltage-transfer characteristic of a common-emitter connected transistor.

current will flow; there will be no voltage drop across R_L, and the collector potential will be approximately equal to the supply voltage V_{CC}. This is the point A on the load line, and the transistor is said to be *cut off.* Now if V_B is made positive and increased, base current flows, collector current flows, and a voltage drop is produced across R_L. The collector voltage V_C moves along the load line as the collector current increases. Eventually, if the base voltage is increased sufficiently, the point B will be reached. The collector voltage is almost 0, and the transistor is said to be *saturated.* The operation can be shown another way by plotting collector voltage (the output voltage) against base voltage (the input voltage), as shown in Fig. 4.10. The points A and B correspond to those in Fig. 4.9b. If the base voltage is changed from a low value X to a sufficiently high positive value Y, the collector voltage will change from a high value V_{CC} to a low value called $V_{CE(sat)}$, the collector-to-emitter voltage with the transistor saturated. When the transistor is used in logic circuits, the condition with the voltage near ground potential is called the logical 0 state, and the condition with the voltage high is called the logical 1 state, as indicated in Fig. 4.10. With the state 1 as the more positive level, as above, it is called *positive logic.* It is also possible to design and use circuits in which the 1 state is a high negative level. This is called *negative logic*—the 0 state is near ground potential, and the 1 state is several volts negative.

The important point to observe in Fig. 4.10 is that the actual values of base input voltage at X and Y can vary somewhat with no resulting change in the output voltage, and so the transistor gives a stable output voltage in both states.

INVERTER CIRCUIT

In the discussion of logic circuits, the NOT function was discussed, in which the output is always in the opposite state to that of the input. The basic transistor circuit discussed above gives this condition and is called an *inverter circuit.* Fig. 4.11 shows

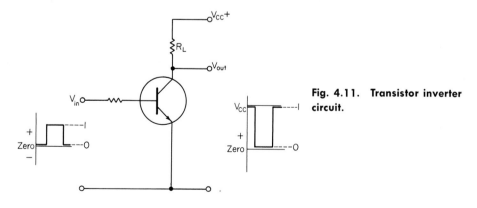

Fig. 4.11. Transistor inverter circuit.

the basic circuit together with the input and output signals. Initially, the input voltage is near 0, state 0, and the output voltage is full positive, state 1. When the input voltage is pulsed up to positive, state 1, the output voltage drops to almost ground potential, state 0.

Although a transistor stage can normally be arranged to give voltage amplification between input and output, this is not generally required in logic circuits, as the full change of output voltage between the 0 and 1 states is available as the input to the next circuit. However, when there is a large fan-out, when the output from a circuit must drive several following circuits, the available input voltage for each of the following circuits may be low. In such cases inverter circuits can be used to provide voltage amplification as well as inversion, to restore the full voltage swing.

BASIC TRANSISTOR LOGIC CIRCUITS

The basic transistor binary circuit can be adapted to give the various logic functions described earlier. There are many ways in which the circuit detail can be arranged. In this section, basic OR, NOR, AND, NAND, and flip-flop circuits will be presented to illustrate the principles. In the next lesson, more detailed alternative circuits will be described to illustrate the present trends of commercially available integrated circuits.

Basic OR and NOR Gates. A basic transistor OR gate using positive logic is shown in Fig. 4.12*a*. It uses a separate transistor for each input, and the emitters of the transistors are connected to a common load resistor R_L. If any one input is pulsed positively from 0 to 1, collector current flows, and the emitter current flowing through R_L will produce a positive output pulse going from 0 to 1. The input and output waveforms are shown on the diagram.

A NOR gate can be made by putting the load resistor R_L in the collector circuit of the transistors to give inversion, as shown in Fig. 4.12*b*. If the input goes positive 0 to 1, the output will voltage drop from 1 to 0.

Basic AND and NAND Gates. A transistor NAND gate is shown in Fig. 4.13*a*. This is called a series NAND circuit. For the transistors to pass a current, all inputs must go positive, 0 to 1, at the same time. Since the load resistor is in the collector, the output voltage will change from V_{CC} (the 1 state) down to the saturation voltage of the three transistors in series (the 0 state), giving the inversion required for the NAND function.

The above circuit can be converted to an AND gate by including a simple inverter stage to invert the signal and give a 1 state at the output for a 1 state at the input. This is illustrated in Fig. 4.13*b*.

Flip-flop Circuit. With all of the circuits described above, it is necessary to maintain the input signal present if it is required to keep the output in a given state. In most cases this is not convenient, and a circuit called a *flip-flop* can be used in which the output remains in a given logical state after the input signal has been removed. The basic flip-flop circuit, shown in Fig. 4.14, consists of two inverter stages, cross-connected between collector and base. If transistor T_1 is initially at zero base voltage and a positive input signal is applied, collector current will flow,

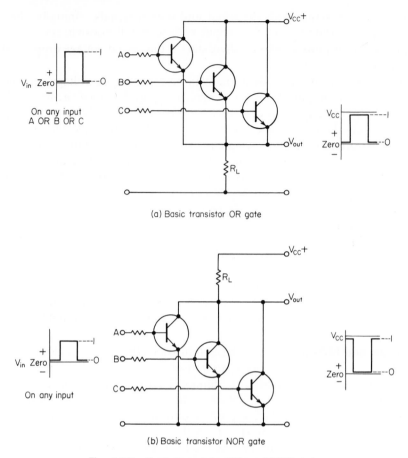

(a) Basic transistor OR gate

(b) Basic transistor NOR gate

Fig. 4.12. Basic transistor OR and NOR gates.

and the collector potential will fall almost to ground potential, to state 0. The base of transistor T_2 is connected to the collector of T_1, and so it will drop to low voltage, T_2 will become nonconducting, and its collector voltage will go positive to the supply voltage, to state 1. The collector of T_2 is connected back to the base of T_1 and will hold it positive when the input signal is removed. The circuit can be reset to its original state by applying a positive pulse to T_2 at R.

THE USE OF THE MOS TRANSISTOR IN LOGIC CIRCUITS

The MOS transistor was described in Lessons 2 and 3, and it was mentioned that it has possible advantages in some aspects; for example, it has a high input resistance and can be fabricated in a very small area. Against this, its operating speed in circuits is slower than that of the bipolar transistor. The MOS transistor can operate as a binary switch in a manner similar to that of the bipolar transistor described above. The p-channel MOS in which a p-type conducting layer is enhanced under the gate electrode is nonconducting for zero gate voltage and so is very convenient

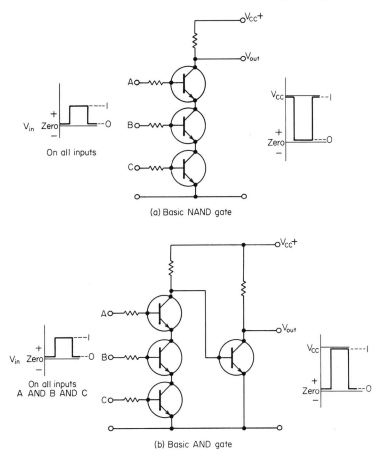

(a) Basic NAND gate

(b) Basic AND gate

Fig. 4.13.   Basic transistor AND and NAND gates.

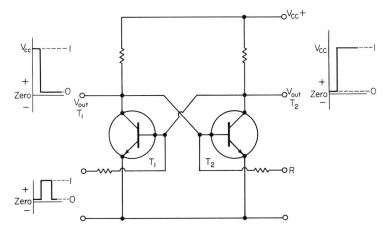

Fig. 4.14.   Basic flip-flop circuit.

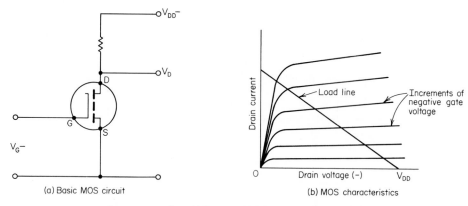

(a) Basic MOS circuit   (b) MOS characteristics

Fig. 4.15. The MOS transistor as a binary switch.

for switching circuits, since one state is defined by zero gate voltage. The drain-supply voltage for this type is negative, so negative logic is used.

The basic circuit using an MOS transistor is shown in Fig. 4.15a, and the output characteristics with the load line are shown in Fig. 4.15b. The similarity to those of the bipolar transistor shown in Fig. 4.9 can be seen.

It was stated in Lesson 3 that the MOS transistor can be used as a resistor between drain and source, the value depending on the dimensions of the structure and the width of the channel formed under the gate. A convenient arrangement is to connect the gate to the drain so that the device will operate in the saturation region and to use dimensions that will give a high value of on resistance. Figure 4.16 shows an MOS inverter circuit using an MOS device as the load resistor. With no signal voltage applied, the driver transistor will be at zero gate voltage and so will be nonconducting. The load device will be conducting, since its gate is connected to

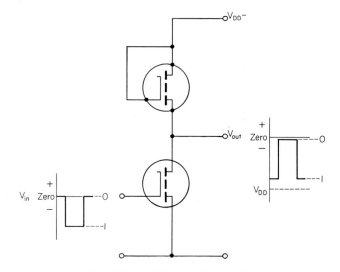

Fig. 4.16. MOS inverter circuit.

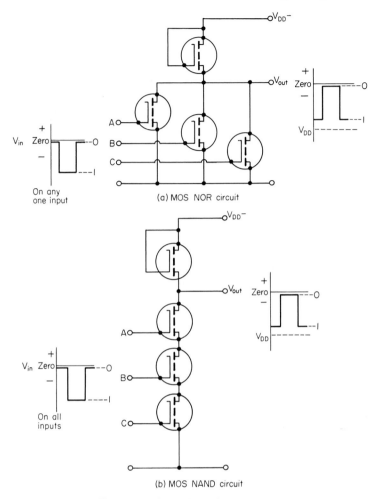

(a) MOS NOR circuit

(b) MOS NAND circuit

Fig. 4.17. Basic MOS logic circuits.

the drain. Thus the output voltage V_D will approach the supply voltage V_{DD}. If the input is pulsed negative, 0 to 1 (negative logic), the driver transistor conducts with a low value of resistance between drain and source, and so the output voltage V_D drops almost to ground potential, going from state 1 to state 0.

MOS versions of the various logic gates are readily possible, and it is usual in each case to use MOS load devices. The basic MOS logic circuits are the NOR and NAND gates as shown in Fig. 4.17. Two inverter circuits can be cross-connected to form a flip-flop circuit as shown in Fig. 4.18. The similarity of these circuits to the bipolar circuits described earlier is readily evident. In checking through their operation, remember that with the p-channel MOS transistor we are using a negative supply voltage and negative logic.

As these MOS circuits use only interconnected MOS structures, they are very suitable for fabrication in integrated-circuit form; and as the area required by MOS

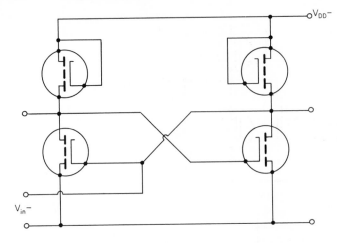

Fig. 4.18.   MOS flip-flop circuit.

structure is very small, quite complex combinations of circuits are possible on a small wafer of silicon.   Thus, MOS integrated logic circuits are very attractive from an economic viewpoint, but as mentioned above, the operating speed of MOS circuits is somewhat limited by capacitance charging and discharging times.   For logic circuits where very fast operating speed is required, the bipolar circuits will offer advantage.

GLOSSARY

analog computer   A computer system in which both the input and output consist of continuously varying signals.

AND gate   A binary circuit having two or more inputs and a single output in which the output is ON (1) only if all inputs are ON (1) together and is OFF (0) if any one of the inputs is OFF (0).

binary system   A system of mathematical computation based on the scale of 2.   A system in which all stages can only have one of two possible states.

cut off   The condition when the emitter junction of a transistor is at 0 or reverse-bias so that no collector current flows.

digital computer   A computer system in which the circuits operate at certain specific signal levels.   In a binary digital computer, the circuits operate with two signal levels, one of which is 0 or very near, and the other is at a defined voltage.

fan-in   The number of inputs to a logic gate.

fan-out   The number of subsequent circuits that a logic gate can drive.

flip-flop   A circuit having two stable states, arranged so that it can be triggered by an input pulse from one state to the other.   In the absence of a trigger signal, the circuit will stay permanently in the state which exists.

full adder   A circuit that will accept three binary-digit inputs (two digits to be added plus a carry digit from a previous stage) and give an output equal to their sum.

gate   A logic circuit having two or more inputs and a single output designed to give an output signal only when a certain combination of input signals exists.

half adder   A circuit that will accept two binary-digit inputs and give an output equal to their sum (including a carry output).

inverter circuit   See *NOT circuit.*

load line   A line drawn on the family of collector characteristic curves of a transistor show-ing how the transistor collector voltage changes as the current flowing through the transistor and load resistance changes.

logic   The basic principles together with circuit and system arrangements involved in mathe-matical computations.

logical 0 state   The state when the logical voltage is at 0 or very near.

logical 1 state   The state when the logical voltage is at a defined relatively high volt-age.

NAND gate   A combination of a NOT and an AND circuit. A binary circuit having two or more inputs and a single output, in which the output is OFF (0) only if all inputs are ON (1) together and is ON (1) if any one of the inputs is OFF (0).

negative logic   A system of logic in which the logical 1 state is characterized by a defined negative voltage.

NOR gate   A combination of a NOT and an OR circuit. A binary circuit having two or more inputs and a single output, in which the output is OFF (0) if any one of the inputs is ON (1) and is ON (1) only if all inputs are OFF (0) together.

NOT circuit   A binary circuit having a single input and a single output, in which the output is always the opposite of the input. When the input is ON (1), the output is OFF (0), and vice versa. This circuit is also called an *inverter circuit*.

OR gate   A binary circuit having two or more inputs and a single output, in which the output is ON (1) if any one of the inputs is ON (1) and is OFF (0) only if all inputs are OFF (0) together.

positive logic   A system of logic when the logical 1 state is characterized by a defined positive voltage.

saturation   The condition when a further increase of a variable has no further increase of the resultant effect. A transistor is said to be saturated when further increase of base current causes no further increase in collector current.

REVIEW

For each of the numbered statements below, select the one of the items lettered *a, b, c,* or *d* that correctly completes the statement.

4.1. Digital computers generally use
    a. The decimal system.
    b. Continuously varying input and output signals.
    c. The binary system.
    d. Vacuum tubes.

4.2. The decimal number 27, expressed in binary form, is
    a. 10111.
    b. 10101.
    c. 11011.
    d. 11101.

4.3. The binary number 1011 is equivalent to the decimal number
    a. 3.
    b. 22.
    c. 19.
    d. 11.

4.4. An OR gate gives a logical 1 output
    a. When any one input is logical 1.
    b. When all inputs are logical 0.
    c. Only when all inputs are logical 1.
    d. Only when any two inputs are logical 1.

4.5. The fan-out of a logic gate is
    a. A system of forced-air cooling.
    b. The number of subsequent circuits the gate can drive.
    c. The number of connections to the package.
    d. The number of inputs to the circuit.

4.6. The NOT circuit
    a. Has several inputs and a single output.
    b. Is the same as a NOR circuit.
    c. Has identical input and output signals.
    d. Is alternatively called an inverter circuit.

4.7. A full-adder system
    a. Needs only two AND gates.
    b. Has only two inputs.
    c. Can be made by combining two half-adder circuits.
    d. Can add any number of "bits" of information.

4.8. In the common-emitter transistor circuit with a collector load resistor, if the base voltage is increased sufficiently
    a. The transistor will cut off.
    b. The transistor will become saturated.
    c. The collector voltage will increase.
    d. The base current will fall to 0.

4.9. The flip-flop circuit
    a. Has unstable output states.
    b. Has two stable states of operation.
    c. Cannot be made with transistors.
    d. Cannot be reset.

4.10. The MOS transistor
    a. Cannot be used for logic circuits.
    b. Can be used for high-speed logic circuits.
    c. Can be used as a load resistor.
    d. Is not suitable for fabrication in integrated-circuit form.

Digital Integrated Circuits

INTRODUCTION

In the previous lesson, it was emphasized that a logic system to carry out a desired function can be designed using "black boxes" which have defined input and output conditions. From the viewpoint of the logic system, it does not matter what is in a box, only what it does. Integrated circuits have become the accepted choice for the logic circuits inside the boxes, but there are many alternative ways in which they can be designed to carry out the various logic functions. Different manufacturers have adopted different approaches to circuit design and a series of competitive logic-circuit families have appeared on the market, with different companies specializing in their preferred designs. The publicity presented on these families of logic circuits, advertised as RTL, ECL, DCTL, DTL, TTL, etc., may have caused some bewilderment and confusion to those not intimately concerned with integrated circuits, but the situation is not really different from the various automobile manufacturers presenting their different types of vehicles, all designed to perform the same overall function.

In this lesson, the "classic" families of integrated logic circuits will be described, to help the nonexpert engineer and technician to become more familiar with their operation and with the differences between the several types. No attempt will be made to present detailed specific information on any of the types. If the basic design and operation are understood, the detail of a particular type is best obtained by perusal of manufacturer's data and, if necessary, discussions with his technical staff.

Early types of digital integrated circuits were based on discrete-component circuit designs, and so they were not necessarily the best arrangement. They were followed by better types, designed to utilize the capabilities and economical considerations of monolithic integrated-circuit techniques. Although some of the earlier types are now tending to drop out, they will be included in this review in order to present a complete evolutionary picture.

Most digital integrated circuits operate between cutoff and saturation, and the various types differ mainly in the method of coupling between inverter stages to give NOR or NAND circuits with the required fan-in and fan-out.

CHARACTERISTICS OF INTEGRATED-CIRCUIT LOGIC GATES

Before going on to describe the various families of integrated-circuit logic gates, it will be appropriate to look at their general characteristics, so that the differences between the types can be better appreciated.

Logic Voltage Levels. Logic circuits are normally connected in cascade; that is, the output from one gate is connected to the input of one or more subsequent gates, and so on. Thus, the switching behavior of one circuit will not only depend on its own output characteristic, but may also depend on the input characteristic of the next gate. Consider the simple case of one inverter circuit driving another as shown in Fig. 5.1. Assume that the input of the first stage is initially positive (logical 1) and is pulsed down to 0 (logical 0). With the second transistor not connected, the first-stage output voltage at B will swing from $V_{CE(sat)}$, about 0.2 volt (logical 0), up to V_{CC} (logical 1). Now consider what happens if the second stage is connected. When the input voltage to the first transistor falls, its collector voltage will start to rise toward V_{CC}, but when it reaches about 0.7 volt, the base-emitter junction of the second transistor will start to conduct, and current will flow down through R_L into its base. Then the voltage at the collector of transistor T_1 will be held at the base-emitter diode voltage of transistor T_2 and so will not rise above about 0.9 volt. Thus, the voltage swing at the output of transistor T_1 will now be from 0.2 (logical 0) to 0.9 volt (logical 1). The value of R_L is such that the current flowing into the base of transistor T_2 will drive T_2 into saturation, and its collector voltage will fall to $V_{CE(sat)}$ to give a logical 0 at the output of the second stage.

The load resistance R_L and the input characteristic of the base-emitter junction of transistor T_2 together result in a combined load line for transistor T_1 as shown in Fig. 5.2a. The forward-voltage transfer characteristic will now be as shown in Fig. 5.2b. It shows that the output swing of transistor T_1, from 0.2 (logical 0) to 0.9 volt (logical 1), gives satisfactory operating points for the input of transistor T_2, which in turn will give the same swing at its output when it is connected to the input of the next stage. The conditions for satisfactory switching between the two states are as follows: When transistor T_1 is saturated, its collector voltage must be low enough to keep transistor T_2 cut off; and when T_1 is cut off, the base current

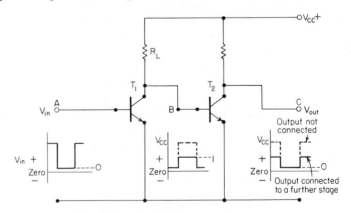

Fig. 5.1. Direct-coupled inverter stages.

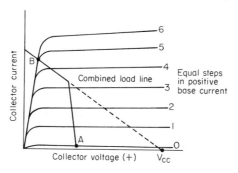

(a) Output characteristics with combined load line

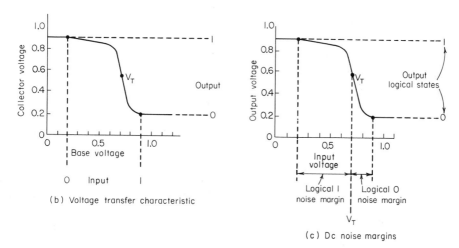

(b) Voltage transfer characteristic

(c) Dc noise margins

Fig. 5.2. Direct-coupled circuit characteristics.

flowing into T_2 must be high enough to saturate T_2. The two operating points with this simple direct-connected circuit are well defined, the 0 state by the collector saturation voltage of the transistor $V_{CE(\text{sat})}$ and the 1 state by the base-emitter diode voltage of the transistor in the saturated condition $V_{BE(\text{sat})}$.

Threshold Voltage. The voltage level at the input of a circuit, at which the circuit changes from one state to the other, is called the *threshold voltage*. One approximation of this is the voltage at the midpoint of the transition between the two states. This is the point V_T in Fig. 5.2b, the threshold voltage in this example being 0.7 volt.

In many logic circuits, the value of threshold voltage is partly dependent on the base-emitter diode characteristics, and since these change with temperature, the threshold voltage will also change with temperature. The detailed circuit design must allow for this. It will be seen later how the value of threshold voltage also depends upon the circuit arrangement.

Noise Margin. In logic systems, the word *noise* refers to any unwanted voltage (dc or ac) appearing at the input of a logic circuit. If such a noise voltage is high enough, it could cause the circuit to change state with no change in signal voltage and so result in false operation of the circuit. The difference between the operating input-logic-voltage level and the threshold voltage is called the *noise margin* of the circuit, that is, the maximum value of noise voltage that the circuit can tolerate without changing states.

In practice, there will be two values of noise margin associated with a circuit: one, the difference between input voltage to give logical 0 output and the threshold voltage and, the other, the difference between the input voltage for logical 1 output and the threshold voltage. These are not necessarily equal as indicated in Fig. 5.2c which shows the case for the direct-coupled circuit of Fig. 5.1. In this case, the noise margin for the circuit must be specified as the smaller of the two noise margins. It will be seen later that recent logic circuits have been designed to have approximately equal logical 0 and logical 1 noise margins, in order to give the most stable operation for a given total logic-voltage swing.

The operating voltages for the logical 0 and logical 1 levels are usually dependent on temperature to some degree, and as mentioned above, the threshold voltage can also vary with change in temperature. The specified value of noise margin must allow for such changes and is quoted as the lowest voltage between the logic operating voltages and the threshold voltage over the full operating temperature range.

It is relatively straightforward to specify the dc noise margin of a circuit, but this is not the whole story. Noise can appear at the input to a logic circuit as short, transient voltage pulses. The actual peak voltage appearing at the transistor base terminal will depend on capacitances present in the circuit and may be different from the corresponding level of a dc noise voltage. The ability of a circuit to withstand such transient noise voltages is called the *ac noise margin* or sometimes the *ac noise immunity*. Since this ac value may be affected by stray capacitances, it will relate to a specific layout arrangement, and moreover, it will depend on the shape and duration of the voltage pulses. Thus, ac noise margin is much more difficult to define and specify.

Actual values of noise margin for various logic circuits will be discussed later in the course.

Operating Speed.   It takes a finite time for a circuit to change from one logical state to the other. In a transistor, this time is that necessary for the base current to supply charge to, or remove charge from, the capacitive elements associated with the transistor structure, in order to produce the required voltage change at the output. In a circuit, additional time will be required to charge any capacitance associated with the load. Thus there is a time delay between the application of a signal at the input, and the change of state at the output. This time delay is called the *propagation delay* of a circuit. Values of delay vary considerably depending on the particular circuit and transistor structure, but with integrated logic circuits they are usually in the range of 2 to 50 nanosec per gate (a nanosecond being 10^{-9} sec). In a complete logic system, there will be a number of gates connected in series, and the overall propagation delay will be the delay per gate multiplied by the number of gates in series.

An electric current propagates along a wire at a rate of about 9 in. per nanosec. With discrete-component circuits, the time delay due to the current traveling along the interconnecting wires could be a significant part of the total circuit delay. With integrated circuits, the interconnections are very short, and the time delay due to current propagation is usually negligible compared with the circuit delay.

Fan-in and Fan-out. The *fan-in* of a logic gate is the number of inputs it is designed to have. This is simply a matter of arranging that the circuit has the required number of logic elements, each with an associated input terminal. A logic gate can be used with any number of inputs up to its maximum fan-in.

The *fan-out* of a logic gate is the number of subsequent circuits that it is capable of simultaneously driving from one logical voltage to the other. The subsequent circuits are all connected in parallel to the single output terminal of the driving gate, and so their input impedances all appear in parallel across the output of the gate. The driving-gate circuit must be designed so that its internal output impedance is low enough to allow the full logic-voltage swing to be produced across the low impedance resulting from the subsequent circuits all being connected in parallel. Thus, the fan-out of a logic gate is specified as the maximum number of circuits that can be connected to its output with satisfactory operation. The gate will operate satisfactorily with any number of circuits less than the specified maximum.

Operating Temperature. All semiconductor devices are temperature sensitive, due mainly to the characteristics of the p-n junctions changing with temperature. Silicon transistors and diodes can operate satisfactorily with junction temperatures up to about 200°C: above this temperature the characteristics become so poor as to result in inferior performance. Diffused resistors may also change their value with change of temperature due to the thermal generation of carriers and change of the mobility of the carriers.

The internal power dissipation of an integrated circuit will cause the junction temperature of the elements to be higher than the ambient temperature. The integrated-circuit manufacturer makes due allowance for this in his design and then specifies the maximum allowable ambient temperature at a value that ensures that there is no possibility of the junction temperature going too high.

Integrated circuits intended for military applications are usually specified to operate over the ambient range of -55 to $+125°$C, and those for industrial use from $0°$ to $+70°$C.

Power Dissipation. The power dissipation of a logic circuit is usually defined as the supply power required for the gate to operate with a 50 percent duty cycle, that is, equal times in the 0 and 1 states. The power dissipation of typical logic integrated circuits ranges from a few milliwatts to about 50 mw per gate, depending on the type of circuit. In general, high-speed circuits with short propagation delay require higher power.

DIGITAL LOGIC-CIRCUIT FAMILIES

As stated earlier, several logic-circuit groups or families have been introduced, based mainly on different methods carrying out the logic and coupling to inverter stages. For example, in direct-coupled transistor logic (DCTL), transistors are used for the logic with direct coupling between stages. With resistor-transistor logic (RTL), series resistors are included in a DCTL circuit. Diode-transistor logic (DTL) uses diodes as logic elements, and transistor-transistor logic (TTL) uses a multi-emitter transistor instead of the diodes. In emitter-coupled logic (ECL), the circuits are coupled by a common-emitter resistor, and complementary transistor logic (CTL) uses a combination of p-n-p and n-p-n transistors. For each family, variations of a basic gate circuit are used to design a range of logic circuits with compatible input and output logical levels. In the design of a complete logic system, it is generally necessary to use logic circuits of one family only.

The several logic families mentioned above will be described in turn, and their relative characteristics, advantages, and disadvantages will be emphasized.

Direct-coupled Transistor Logic (DCTL). In the DCTL system, the output of a gate is directly connected to the input of the next gate. Figure 5.3 shows a DCTL NOR gate. The input voltage is normally derived from the collector of the previous gate, and the output connects directly to input of the following gate, as indicated by the dashed circuits.

If a positive voltage (logical 1) is connected to input *A* or *B* or *C,* the respective transistor will saturate, and the output voltage will drop to the saturation voltage of the transistor to give a logical 0 output. With the dashed driving and load gates connected, the logical voltage swing at both input and output of the NOR gate will be approximately from 0.2 (logical 0) to 0.9 volt (logical 1), as previously described with reference to Fig. 5.2. The threshold voltage will be about 0.7 volt. The advantage of the system is its simplicity, but it has the problem that its operation is affected by slight differences between the characteristics of different transistors. If one transistor has a base-emitter voltage slightly lower than others in parallel, it takes most of the available current and prevents proper overall operation of the circuit. This is called *current hogging.* To reduce the effect, resistors are included in series with each base lead, so that the base current is less dependent on the individual base-emitter characteristics. The circuit is then known as *resistor-*

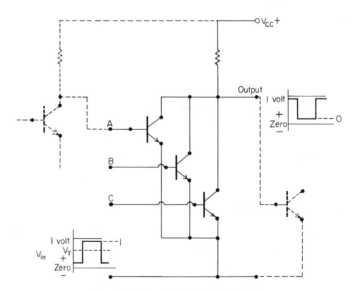

Fig. 5.3. Basic DCTL NOR circuit.

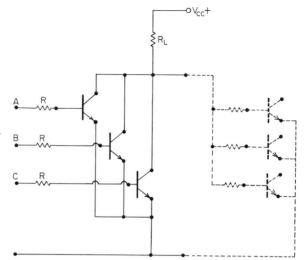

Fig. 5.4. Basic RTL NOR circuit.

transistor logic. The simple DCTL circuit is now rarely used and will not be discussed further.

Resistor-transistor Logic (RTL). Resistor-transistor logic was the first family of logic circuits established as a standard catalog line. The basic arrangement is shown in Fig. 5.4, which shows the series resistors added to each transistor. By reducing the current-hogging effect, the use of the resistors allows a larger fan-out. Against this, the series resistors have an adverse effect on the switching speed of the circuit, since the input capacitance of the transistors must now be charged and discharged through additional resistance which gives the circuit an increased time constant. Thus, with RTL, there must be a compromise between fan-out and switching speed. Typical values are a fan-out of 4 or 5 with a switching delay of 50 nanosec. The operating points and logical voltage swing are similar to those for the DCTL circuit. The RTL circuit has a relatively poor noise immunity. The noise margin from the logical 0 state to the threshold voltage is about 0.5 volt, but from the logical 1 state to the threshold voltage it is only 0.2 volt.

The switching speed of the RTL circuit can be improved by the addition of a capacitor in parallel with the series resistor. This variation is called *resistor-capacitor-transistor logic* (RCTL) and is shown in Fig. 5.5. The capacitor allows the leading and trailing edges of a signal pulse to bypass the resistor so that the transistor input capacitance charges up quicker. The use of the capacitor also allows higher values of resistor, with the possibility of lower power dissipation per gate. The RCTL circuit, however, is not ideal from the integrated-circuit viewpoint, as it includes a high proportion of resistors and capacitors, which, as discussed in Lesson 3, are relatively costly because of the large area they occupy. RTL and RCTL circuits are still used in established equipment, but they are rarely used for any new development.

Diode-transistor Logic (DTL). The next family of integrated logic circuits to become established was the *diode-transistor logic* circuit shown in Fig. 5.6. The logic is performed by the input diodes D_1, D_2, and D_3, and the signal is then coupled

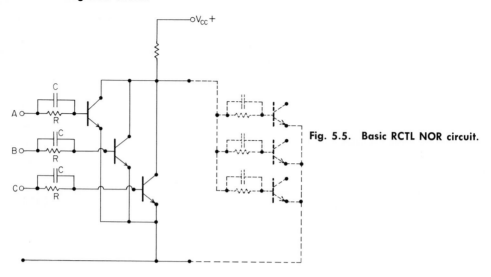

Fig. 5.5.   Basic RCTL NOR circuit.

through a series diode D_S to an inverter stage consisting of a transistor and its load resistor.   The overall DTL circuit constitutes a NAND gate.

If all inputs are at logical 1 with a positive signal voltage equal to V_{CC}, the three input diodes will be reverse-biased and so will pass no current.   The series diode D_S is forward-biased, and so current flows through R_D and D_S into the base of the transistor and holds it in saturation, with the low collector voltage $V_{CE(sat)}$ giving a logical 0 at the output.

If any one of the inputs drops to ground potential, logical 0, the corresponding input diode conducts, and current flows down through R_D and the diode.   The potential at the point X drops to the voltage across the input diode which will now be about 0.7 volt, and this is not sufficient to drive current through the diode D_S plus the base-emitter junction of the transistor in series.   Thus no base current flows, the transistor is cut off, and the collector potential rises to V_{CC} to give a logical 1 output.

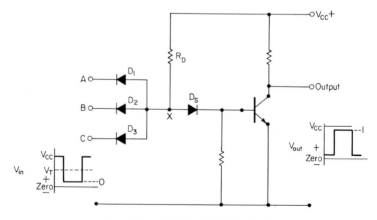

Fig. 5.6.   Basic DTL NAND circuit.

Going back to the condition when all inputs are at positive V_{CC} potential, the input diodes are reverse-biased, and current flows through R_D and D_S into the base of the transistor. The potential at point X will be approximately 1.4 volts, 0.7 volt across the series diode D_S and 0.7 volt across the base-emitter junction of the transistor. Now let one input voltage be reduced gradually. For the associated input diode to start to conduct, it is necessary to reduce the input voltage down to 0.7 volt so that there is a forward voltage of 0.7 volt across the diode. Then the diode conducts, the voltage at point X falls to 0.7 volt, and the transistor is cut off. Thus it will be seen that the threshold voltage of this circuit is 0.7 volt. If two diodes in series are used for D_S, the threshold voltage is increased by an additional 0.7 volt to a value of 1.4 volts; this is usually done in practical DTL circuits.

With a logical 1 input, the input resistance of the gate is high—the diodes are reverse-biased, and so the gate does not load the previous circuit. Thus the logical 1 output level from a previous DTL circuit can be the full supply voltage V_{CC}. If this is set at 4 volts, the two operating points at input and output of the gate will be 0.2 volt for logical 0 and 4 volts for logical 1. If two series diodes are used with a threshold voltage of 1.4 volts, the 0-state noise margin will be 1.2 volts which is substantially better than that of the RCTL circuit.

The DTL circuit switches faster than the RTL circuit, since the signal passes through the low forward resistance of the diodes to the transistor. A typical delay time is 25 nanosec. A high fan-out up to the order of 8 is possible because of the high input impedance of the subsequent gates in the logical 1 state. The use of diodes in the DTL gate, rather than resistors and capacitors as in RCTL, makes it more economical in integrated-circuit form.

Transistor-transistor Logic (TTL). In the transistor-transistor logic circuit, a single multiemitter transistor replaces the input diodes and the series diode as shown in Fig. 5.7. Each emitter-base diode serves as one input, and the base-collector diode operates as the series diode.

The operation of the circuit is basically similar to that of the DTL circuit. Suppose that all inputs are logical 1 at V_{CC} potential. All the emitter junctions of transistor

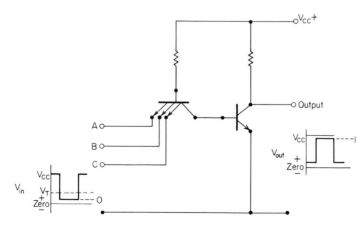

Fig. 5.7. Basic TTL circuit.

T_1 are reverse-biased, but the collector junction is forward-biased, and current flows through it to drive transistor T_2 into saturation and give a logical 0 output.

Now let any one of the inputs be logical 0 near ground potential. The emitter of transistor T_1 connected to that input is then forward-biased, and current flows through it and causes the potential at the base of T_1 to fall; transistor T_2 is cut off, and the collector voltage rises to V_{CC} to give a logical 1 output. The conditions are similar to those existing with the DTL circuit of Fig. 5.6.

The multiple-emitter transistor is economically fabricated in monolithic form. A single isolated collector region is diffused, a single base region is formed in the collector region, and then the several emitter regions are diffused as separate areas into the base region.

This logic circuit is used as the basis for a number of TTL logic gates. A full NAND gate is shown in Fig. 5.8a. An output stage is added to the basic logic circuit to give current gain drive for switching in both directions, resulting in faster switching speed and a higher fan-out capability. Since the base-emitter junction of transistor T_4 is in series with that of transistor T_2, the threshold voltage is increased to about 1.4 volts, giving a noise margin of the order of 1.2 volts.

A TTL NOR gate is shown in Fig. 5.8b. The input logic circuit of T_1 and T_2 is duplicated by T_3 and T_4 with only one emitter input each to T_1 and T_3. The outputs of T_2 and T_4 are connected in parallel and fed to an output stage. If input A is at V_{CC} potential (logical 1), T_1 emitter-base junction will be cut off, and current will flow through the 4-kilokm resistor to drive T_2 into saturation. Alternatively, if input B is at V_{CC}, T_4 will be driven into saturation. Thus if either input A OR input B is positive, at logical 1, current will flow through the common 1.6-kilokm load resistor and the common 1-kilokm emitter-resistor of transistors T_2 and T_4 to drive the output stage and give NOR operation.

Figure 5.8c shows a TTL NAND gate which has a power output stage to give a very low output impedance and allow a high fan-out up to about 30. By including an additional inverter stage, the circuit can be converted to an AND gate as shown in Fig. 5.8d.

The TTL circuit is very adaptable to all forms of integrated-circuit logic and gives medium-cost circuits with high noise margin, high operating speed, high fan-out, and reasonably low power consumption.

Emitter-coupled Logic (ECL).   A basic emitter-coupled logic gate is shown in Fig. 5.9. The emitters of the logic transistors T_1, T_2, and T_3 are coupled to the emitter of a reference transistor T_4. The common-emitter resistor R_E is high enough to act as a constant current source. The base of transistor T_4 is connected to a reference voltage V_{BB}. If the inputs are all near ground potential, logical 0, transistors T_1, T_2, and T_3 will all be cut off, no current flows through R_1, and the common-collector potential rises towards V_{CC}. This drives transistor T_5 into conduction, and the output from the emitter of T_5 goes positive to give a logical 1 output.

If one of the inputs is made positive, above the value of V_{BB}, to give a logical 1 input, current will flow through the associated transistor, causing the collector potential and the output voltage from T_5 to fall to give a logical 0 output. Since resistance R_E constitutes a constant current source, as the current through the logic transistor increases, the current through the reference transistor decreases. The

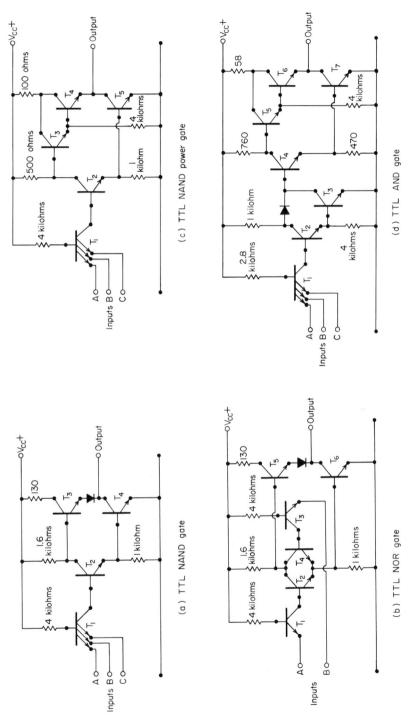

Fig. 5.8. Alternative TTL gate circuits.

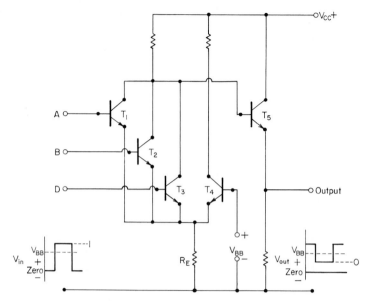

Fig. 5.9. Basic ECL NOR circuit.

switching threshold voltage is equal to the reference voltage V_{BB}. The use of emitter coupling in both the logic stage and output stage prevents the transistors from going into saturation, and as a result, the switching speed is very fast, typically only a few nanoseconds. The power dissipation, however, is relatively high, typically 50 mw. The input threshold voltage, being equal to the reference voltage V_{BB}, is well defined, but the logical 0 and 1 levels cannot be defined as precisely as in saturated circuits. The use of an emitter-follower circuit for the output gives a very low output impedance, allowing a very high fan-out up to the 20 to 30 region.

Complementary-transistor Logic (CTL).   This is a family of logic circuits using an emitter-coupled OR gate with a combination of p-n-p and n-p-n transistors. A typical example is shown in Fig. 5.10. With all inputs at V_{CC}, logical 1, the input transistors (p-n-p) are all cut off, no current flows, and the common-emitter voltage rises towards V_{CC}. Base current flows into transistor T_4 (n-p-n), and the transistor conducts to give a positive output voltage, logical 1. If one of the inputs is reduced to ground potential, logical 0, the associated p-n-p transistor conducts, the current through R_1 causes the common-emitter potential to fall, and the current through T_4 falls, giving a logical 0 ouput. As with the ECL circuit, the transistors do not go into saturation, and so switching speed is fast. The emitter-follower output gives a high fan-out capability, and the high input impedance allows easy coupling into the circuit.

MOS Logic Circuits.   The basic MOS logic circuits were described in Lesson 4. To look further into their operation, first consider the basic MOS inverter stage shown in Fig. 5.11. The transfer characteristic of a p-channel MOS transistor is shown in Fig. 5.12a. With MOS transistors, the threshold voltage is defined as the gate voltage at which conduction between drain and source just starts, and is approximately 4 volts. The voltage transfer characteristic of an MOS inverter stage

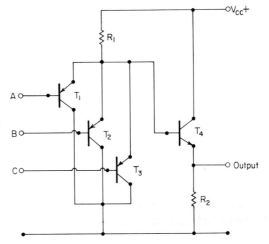

Fig. 5.10. Basic CTL OR circuit.

with an MOS load device is shown in Fig. 5.12b. As mentioned in Lesson 4, with the p-channel MOS transistor, negative logic is used. With zero input voltage, logical 0, the output voltage is the supply voltage V_{DD} minus the threshold voltage of the load device, which is also about 4 volts. Thus with a supply voltage of -15 volts, the output voltage will be -11 volts, which becomes the logical 1 level. If a negative input voltage exceeding the threshold voltage is applied, the circuit conducts, and the output voltage drops to a low value, typically -0.5 volt, which becomes the logical 0 level. Thus the output logical swing will be from -11 to -0.5 volts. If a second inverter stage is added, as shown in Fig. 5.13, the high input impedance of the second stage will not affect the voltage levels existing at the output of the first stage, and the swing between -11 and -0.5 volts will drive the second stage so that its output swings between -0.5 and -11 volts. Typical noise margin between the 0 state and the threshold voltage is about 3.5 volts, but it must be remembered that the gate impedance level is very high, and so only a small unwanted induced energy may generate a significant noise voltage.

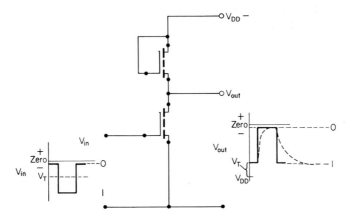

Fig. 5.11. MOS inverter circuit.

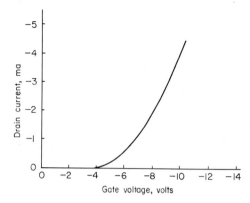

(a) Transfer characteristic of MOS transistor

Fig. 5.12.   MOS transfer characteristics.

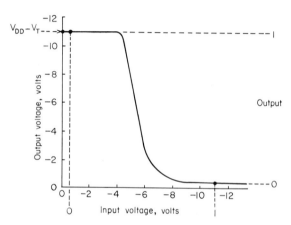

(b) Voltage transfer characteristic of MOS inverter circuit

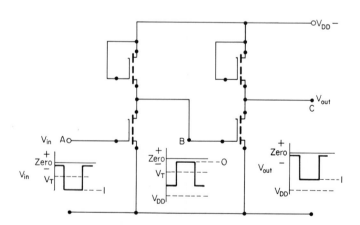

Fig. 5.13.   Direct-coupled MOS inverter stages.

It has been mentioned that MOS circuits are limited in switching speed compared with bipolar transistor circuits. This is due to the fact that the MOS transistor is a higher-impedance device and so cannot charge the stray circuit capacitance quickly. The shape of the switching waveform of an MOS inverter with a very short duration signal pulse is indicated by the dashed shape on the output pulse of Fig. 5.11. With the input at zero voltage, the output voltage is high, and the circuit capacitance, consisting of the stray capacitance of the interconnections and the input capacitance of the next stage, is charged up to the level of -11 volts. When a high negative voltage, logical 1, is applied to the input, the output voltage drops, and the capacitance discharges through the driver device. The ON resistance of an MOS device is much higher than the resistance of a bipolar transistor in the saturation region, and so the discharge time will be relatively long. When the input returns to zero, the capacitance must charge back up through the load device, which has an ON resistance higher than that of the driver, and so the rate of charge is slower than the discharge. Typical switching times at present are a fraction of a microsecond, but advances in circuit design and fabrication technology will doubtless enable this to be reduced somewhat.

With MOS integrated circuits, logic is most conveniently carried out using the NOR circuit illustrated in Fig. 4.17a, as this arrangement gives the lowest value of ON voltage for the logical 0 level.

SUMMARY AND COMPARISONS

Table 5.1 summarizes the properties and features of the various integrated-circuit logic families.

DCTL is the simplest system, but it suffers from current hogging; this is improved in RTL and RCTL by the use of series resistors. All have poor noise margins. DTL uses diodes for the logic, and has higher noise margin and faster switching. TTL uses a multiemitter transistor instead of diodes and can switch faster with a high fan-out and good noise margin. ECL and CTL are nonsaturating circuits and

Table 5.1.   Comparison of Integrated-circuit Logic Families*

Family	Logic type	Relative cost per gate	Propagation time per gate, nanosec	Power dissipation per gate, mw	Typical noise margin, volts	Typical fan-in	Maximum fan-out
DCTL	NOR	Low	15	10	0.2	3	3
RTL	NOR	Medium	50	10	0.2	3	4
RCTL	NOR	Medium/high	30	10	0.2	3	4
DTL	NAND	Medium	25	15	0.7	8	8
TTL	NAND	Medium	10	20	1.0	8	12
ECL	OR/NOR	High	2	50	0.4	5	25
CTL	AND/OR	High	5	50	0.4	5	25
MOS	NOR	Very low	250	<1	2.5	10	5

*Values quoted are representative on a comparison basis. For any one family, the values may vary somewhat.

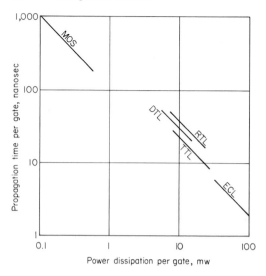

Fig. 5.14. Propagation time versus power spectrum of logic integrated circuits.

have the highest speed and fan-out, but they also have the highest power consumption. MOS circuits are extremely simple and economic, but they are limited to low- and medium-speed applications.

In the detailed design of each family of circuits, there is a compromise between switching speed and power dissipation, which allows variations depending upon the specific requirement for a circuit. Figure 5.14 shows the relation between typical propagation time and power dissipation for the various integrated-circuit logic families.

GLOSSARY

complementary-transistor logic (CTL)   A logic system using emitter-coupled circuits with a combination of p-n-p and n-p-n transistors.

current hogging   A condition when one of several circuits connected in parallel has a lower resistance and takes most of the available current, resulting in unequal current sharing.

diode-transistor logic (DTL)   A logic system in which the logic decisions are carried out by a group of diodes, and the resulting output is coupled through a transistor output stage.

direct-coupled transistor logic (DCTL)   A system of transistor logic in which the collector output of one gate is connected directly to the base input of the next gate.

emitter-coupled logic (ECL)   A logic system using emitter-coupled transistor circuits.

noise margin   The difference between the operating voltage of a binary logic circuit and the threshold voltage.

power dissipation of a logic circuit   The supply power when a logic circuit is operating with a 50 percent duty cycle, that is, equal times in the 0 and 1 states.

propagation delay   The time delay between the application of a signal to the input of a logic circuit and the change of state at the output.

resistor-capacitor-transistor logic (RCTL)   A variation of the resistor-transistor logic system in which a capacitor is connected across the series resistor to allow faster switching.

resistor-transistor logic (RTL)   A system of transistor logic in which a resistor is included in series with the base of each transistor in order to reduce current-hogging effects.

threshold voltage   The input voltage level at which a binary logic circuit changes from one state to the other.

transistor-transistor logic (TTL)   A logic system similar to diode-transistor logic in which the logic diodes are replaced by a multiemitter transistor.

REVIEW

For each of the numbered statements below, select the one of the items lettered *a, b, c,* or *d* that correctly completes the statement.

5.1. One condition for satisfactory switching of a DCTL circuit is that
  a. The transistors must never go into saturation.
  b. When one transistor is saturated, its collector voltage must be low enough to cause the next transistor to be cut off.
  c. The input impedance must be very high.
  d. The transistors must be coupled with a common-base arrangement.

5.2. The threshold voltage of a logic gate is
  a. Independent of temperature.
  b. Not related to noise immunity.
  c. Always the same as the logical 1 voltage.
  d. The voltage at which the circuit changes from one state to the other.

5.3. The propagation delay of an integrated-circuit logic gate is
  a. Typically 5 microseconds.
  b. The time for the signal to travel from the integrated-circuit wafer to the package terminals.
  c. Usually lower than that of the corresponding discrete-component circuit.
  d. Independent of load capacitance.

5.4. Current hogging
  a. Is not a serious effect.
  b. Is due to unequal input characteristics of transistors connected in parallel.
  c. Cannot be reduced by circuit design.
  d. Is a limitation in TTL circuits.

5.5. The use of the series base resistor in RTL circuits
  a. Increases operating speed.
  b. Improves noise immunity.
  c. Reduces current-hogging effects.
  d. Decreases the possible fan-out.

5.6. The capacitor in RCTL circuits
  a. Increases the possible fan-in.
  b. Increases the power consumption.
  c. Increases the operating speed.
  d. Makes the circuit more suited to fabrication in integrated-circuit form.

5.7. The DTL circuit
  a. Has a higher propagation delay than the RTL circuit.
  b. Has a higher noise margin than the RTL circuit.
  c. Has a low fan-out capability.
  d. Is not economical to fabricate in integrated-circuit form.

5.8. The TTL circuit

    a. Uses a multiemitter transistor for the logic decisions.

    b. Has a noise margin about the same as that of the DCTL circuit.

    c. Is only suitable for NAND gates.

    d. Is not adaptable to integrated-circuit logic systems.

5.9. With the ECL circuit,

    a. Power consumption is relatively low.

    b. The transistors operate in the saturated mode.

    c. The threshold voltage cannot be varied.

    d. The switching speed is very fast.

5.10. With a p-channel MOS logic circuit,

    a. Input current is high.

    b. Positive logic is used.

    c. The threshold voltage is about negative 4 volts.

    d. The noise margin is lower than that of the TTL circuit.

LESSON **6**

Basic Aspects of Linear Integrated Circuits

GENERAL CONSIDERATIONS

A linear circuit is one in which a proportional relation is maintained between the input and output signals at all times. The usual function carried out by a linear circuit is amplification. In a linear amplifier, the amplified output signal is directly proportional to the input signal; that is, the output-signal waveform is an amplified replica of the input-signal waveform.

A linear transistor amplifier operates in the linear region of the forward-voltage transfer characteristic as indicated in Fig. 6.1. The transistor is biased so that with no input signal (the quiescent state) the operating point P is near the center of the linear region of the characteristic, to allow the maximum possible linear swing of the signal about the point. The amplitude of the input signal must not extend to the cutoff or saturation regions or distortion of the output signal will result and linear operation will be lost. It is important that the operating point be stabilized to prevent shift due to changes of temperature and supply voltage. This is one of several requirements that have led to the development of new techniques for monolithic linear integrated circuits.

Compared with discrete-device circuits, the design of linear integrated circuits is made difficult by limitations of the passive elements. Diffused resistors cannot be reproduced with a tolerance better than about ± 5 percent, and integrated-circuit capacitors are limited to about 100 pf on economic grounds. Also, it is not possible to form useful inductance in a small wafer of silicon, and this imposes a further limitation with high-frequency tuned circuits; it is necessary to add the tuning elements as external components. Although it is possible to achieve some frequency selection by resistor-capacitor networks, the scope is considerably inferior to that of inductance-capacitance tuning, which is so very straightforward.

Against these difficulties, certain aspects of monolithic integration offer distinct advantage for linear circuits. Additional transistor and diode elements can be included at small extra cost, allowing wider scope in circuit design. Diffused resistors can be processed side by side with a very close ratio of their resistances, down to the order of 1 percent. Because all the elements of a monolithic circuit are formed in the same small wafer of silicon, they will all tend to be at the same temperature

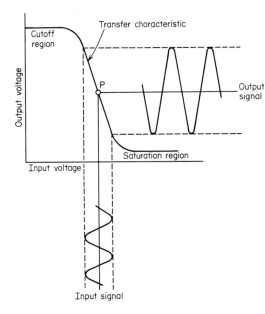

Fig. 6.1.   Linear amplification.

and will track together with change of temperature. All these advantageous points have been utilized in the design of integrated-circuit differential amplifiers which give excellent performance. These will be discussed in some detail later in this lesson.

Linear integrated circuits can be used for any frequency from dc up to several hundred megahertz, with signal-output power levels approaching 1 watt. Higher-power circuits with outputs up to several watts are now being made for audio amplifier applications. Typical applications of linear integrated circuits are in microphone amplifiers, in hearing aids, in communication equipment, in radio and television receivers, and, very important, in operational and differential amplifiers for analog computers and electronic control systems.

The small physical size of monolithic integrated circuits gives the possibility of locating a complete preamplifier within a transducer and so reducing the effect of electrical noise pickup in the leads running to a remote amplifier and resulting in a better overall signal to noise performance. A typical example of this is including a preamplifier within a magnetic microphone.

There is a very important area of application of linear integrated circuits in the microwave region. Here, the small dimensions required for microwave strip-line tuning elements are compatible with integrated-circuit-device dimensions. At present, most microwave integrated circuits are of the hybrid type, using thin-film microwave strip-line circuits with separate transistor and diode chips on ceramic substrates, but there is a possibility that they will eventually be produced in monolithic form. A brief discussion of the present hybrid microwave integrated circuits will be given later in this lesson.

BASIC TRANSISTOR LINEAR AMPLIFIER CIRCUITS

In order to appreciate the points involved in the design of linear integrated circuits, it is necessary to understand the basic operation of linear transistor amplifiers. Consider the simple n-p-n common-emitter amplifier stage shown in Fig. 6.2a. Assume that the bias battery V_{BB} sets the quiescent operating point to a suitable level. If the signal input voltage goes positive by a small amount, the base-emitter junction is biased more in the forward direction, resulting in an increase of base current flowing into the transistor. This increase of base current is amplified by the transistor, and the collector current increases by an amount equal to the change in base current multiplied by the current gain of the transistor. This increase of collector current flowing through the load resistor results in an increase in the voltage drop across the load resistor, and the output (collector) voltage goes down by the same amount.

If the input signal voltage is decreased, the base current decreases, the collector current decreases by an amplified amount, the voltage drop across the load resistor decreases, and the output voltage increases.

Thus, a small change of input voltage causes an amplified change of the output voltage in the opposite sense. We say that the common-emitter stage gives voltage amplification with phase reversal.

An alternative arrangement is the emitter-follower circuit shown in Fig. 6.2b. If the signal input voltage goes positive, the base current increases, and the current gain of the transistor results in an amplified increase in the collector current. This current passes through the load resistor in the emitter circuit to give a positive going output voltage. This output voltage, however, also appears in the input circuit and is such as to oppose the increase in base-emitter voltage of the transistor. As a result, the output voltage will always be slightly less than the input voltage by the small voltage across the base-emitter junction of the transistor. The output voltage "follows" the input voltage and hence the name. Thus, with the emitter-follower circuit, we have a high current gain, a voltage gain just less than unity, and no phase change. A very important characteristic of the circuit is that it has a very low output impedance.

Multistage Circuits. When several amplifier stages are connected in series to obtain a high overall amplification, it is necessary to couple the output of one stage

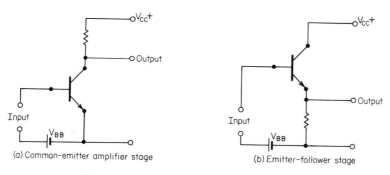

(a) Common-emitter amplifier stage   (b) Emitter-follower stage

Fig. 6.2. Basic transistor amplifier stages.

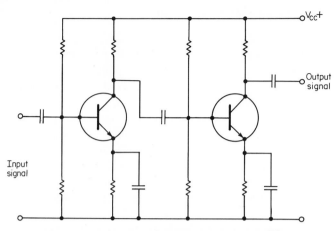

Fig. 6.3.   Two-stage discrete transistor amplifier.

to the input of the next in such a way that the dc operating points of the several stages are correct.   With discrete-device circuits, this is usually accomplished by setting the operating points for each circuit and then coupling the circuits through capacitors large enough to pass the ac signals.   The capacitors give dc isolation, and so the operating point of each stage is maintained.   A typical two-stage amplifier designed to use discrete components is shown in Fig. 6.3.   The resistance network arrangement in the base-emitter circuit is to stabilize the operating point against temperature change.   With monolithic integrated circuits, it is not possible to form capacitors with large values, and as described in Lesson 3, the use of medium- and high-value resistors is uneconomic.   As a result, new circuit techniques have been developed for integrated circuits, to eliminate the use of capacitors and to use transistors or diodes rather than resistors wherever possible.   Direct coupling between stages avoids the use of capacitors as shown in Fig. 6.4a.   The operating point of the second stage is set by the value of load resistor of the first stage and the emitter resistor of the second stage.   An alternative method is to use alternate n-p-n and p-n-p stages as shown in Fig. 6.4b.

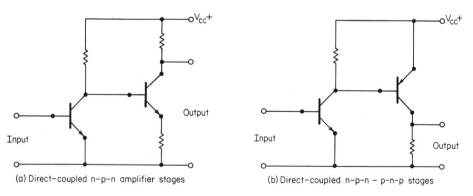

(a) Direct-coupled n–p–n amplifier stages   (b) Direct-coupled n–p–n – p–n–p stages

Fig. 6.4.   Direct-coupled transistor amplifier.

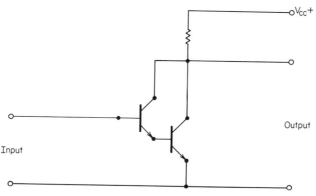

Fig. 6.5.   Darlington-coupled amplifier.

An interesting arrangement often used is the Darlington connection shown in Fig. 6.5. Signal current enters the first transistor, is amplified, and the amplified current is fed directly into the base of the second transistor where it is amplified again. The overall current gain is equal to the product of the separate gains of the two transistors and can be several thousand times. The combination can be readily fabricated in integrated-circuit form and used as a single high-gain stage to considerable advantage in linear integrated circuits.

One of the problems associated with direct-coupled circuits is that of dc drift. This is due to the fact that the operating point of a transistor moves with change of temperature, and even with circuit stabilization, this may result in a small change in the dc output voltage of an amplifier stage. This small dc change will be amplified by subsequent direct-coupled stages and may result in an excessive change in the final dc output level. The main cause of this drift is the variation of the transistor base-emitter junction forward characteristic with temperature. A convenient method of reducing the effect of this variation is the differential-amplifier stage, in which a second transistor is used to balance out any dc changes other than that due to the signal input.

THE DIFFERENTIAL STAGE

The differential-amplifier stage is the main building block of practically all monolithic linear integrated circuits and so is of considerable importance. The basic circuit is shown in Fig. 6.6a. There are two important features of the arrangement. First, the two transistors and the two load resistors are made as identical as possible, and the common emitter resistor has a high value so that it acts as a constant current source. With no signal input to either side, the circuit is balanced. The currents through the two transistors are equal, the two collector voltages are the same, and there is no differential voltage between the two output terminals. If an increase in temperature occurs, any change in current is the same through each transistor, so any voltage change is the same at the two collectors, and there is still no differential output voltage between the output terminals. Moreover, the situation is improved further by the constant-current emitter resistor which tends to maintain the total

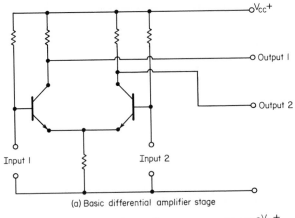

(a) Basic differential amplifier stage

Fig. 6.6.   Differential-amplifier stages.

(b) Differential stage with constant-current transistor

current through both transistors constant, and so it tends to maintain the voltage at each collector constant. If a positive signal is applied to one of the inputs, the current through the transistor on that side increases. Since the total current is constant, the current through the other transistor decreases by an equal amount. The first collector voltage goes less positive, and the second goes more positive. Thus we can take an output signal from either collector, depending on the phase required, or between the two collectors to give a differential output. In either case, the output is proportional to the difference between the signal voltages at the two inputs.

The ability of a differential circuit to reduce the dc drift due to temperature or to balance out a signal which is common to both input terminals is called its *common-mode rejection;* this is an important parameter of a direct-coupled amplifier.

The emitter resistor can be replaced by a transistor to give the constant-current operation as shown in Fig. 6.6*b*. The value of the constant current is set by connecting the V_{BB} terminal to a suitable voltage. The transistor operates on the horizontal part of the collector current versus collector voltage characteristic.

The differential circuit is particularly suitable for integrated circuits. The two amplifier transistors and the two load resistors are formed simultaneously and so will match very well. Because the wafer is small, the two halves of the circuit will always tend to be at the same temperature. The use of a transistor constant-current circuit is advantageous from the integrated-circuit viewpoint, as it is more economic than a high-value resistor and allows the possibility of introducing negative feedback from a subsequent stage to improve the common-mode rejection further.

NEGATIVE FEEDBACK

The amplification of a transistor amplifier will depend upon the current gain of the individual transistors used in the amplifier, on the value of the load resistors, and on the circuit arrangement. It may vary with changes of ambient temperature and supply voltage. In many applications, it is desirable that an amplifier have a known precise value of amplification that will be constant under any conditions. This can be achieved by the use of *negative feedback*.

Consider the arrangement shown in Fig. 6.7. We have an amplifier with a nominal gain of A, and the output is connected to a feedback network which results in a fraction (β) of the output voltage being fed back to the input. The fraction β will equal $r/(R + r)$, and the feedback voltage V_f will equal βV_o. The feedback voltage is connected into the input circuit so that it opposes the input signal voltage V_i. Thus the voltage at the input to the amplifier is now equal to $V_i - \beta V_o$, and this is amplified to give an output $A(V_i - \beta V_o)$.

Thus $V_o = A(V_i - \beta V_o)$, and transposing $V_o (1 + A\beta) = V_i A$. The gain from input to output with the feedback is $V_o/V_i = A/(1 + A\beta)$. If the product $A\beta$ is made large compared with unity, this reduces to $1/\beta$, and the gain is then *dependent only on β*, which is the ratio of two resistances. Thus the overall gain can be set to any precise value by choice of the resistors in the feedback network. The amplifier gain without feedback is called the *open-loop gain* (A_o), and the gain with feedback the *closed-loop gain* (A_c). If we have $A_o = 10,000$ and feedback 1 percent of the output ($\beta = 0.01$), the closed-loop gain will be $10,000/(1 + 100) \approx 99$. If A_o falls to 5,000, the closed-loop gain will only fall to 98.

In addition to giving a constant-gain system, negative feedback can be used to

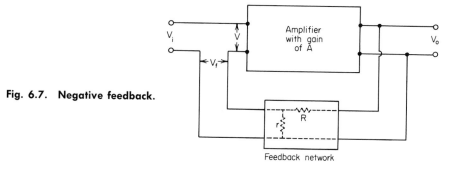

Fig. 6.7. Negative feedback.

Feedback network

Feedback voltage $V_f = \beta V_o = V_o \dfrac{r}{R+r}$

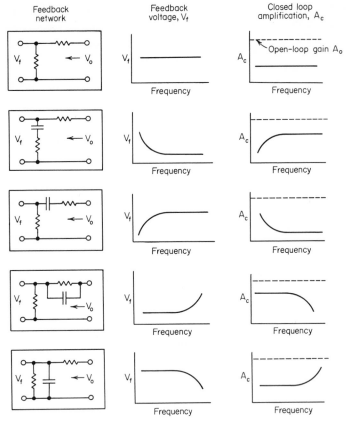

Fig. 6.8. Frequency-response control by negative feedback.

give any required frequency response by including capacitive elements in the feed-back network. The general principle is illustrated in Fig. 6.8. The resistance-capacitance combination in the feedback network is chosen to give a feedback voltage with a frequency response equal to the inverse of that required at the amplifier output. Variation of the feedback voltage can often be used as a convenient means of gain control.

Some negative feedback is incorporated in most linear integrated circuits, and provision is made so that additional feedback can be applied via external circuits to allow a wider application.

A typical example of negative feedback in a simple two-stage common-emitter amplifier is shown in Fig. 6.9. If the input signal goes more positive, the collector of the first stage goes less positive, and the emitter of the second stage also goes less positive. A fraction of the second-stage emitter voltage is fed back to the base of the first stage, and this is such as to oppose the positive signal change.

BASIC TYPES OF LINEAR INTEGRATED CIRCUITS

Linear integrated circuits can be designed to operate over a wide range of frequencies and perform a variety of functions. In general they use combinations of direct coupling, differential stages, and negative feedback to give the characteristics

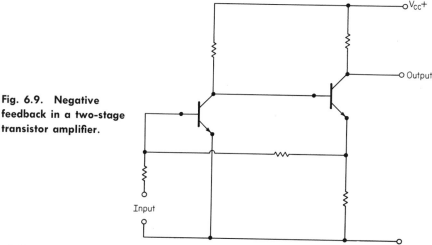

Fig. 6.9. Negative feedback in a two-stage transistor amplifier.

required for each type. In the following paragraphs, a brief description of each main class of linear circuit will be given. The detailed circuit design of specific types is beyond the scope of this course, but the basic design approach and objectives of each type will be covered in order to give a general understanding of some of the important application criteria. In the next lesson (Lesson 7), some details of commercial types will be given to indicate what specific circuits are available and what they will do.

Audio Amplifiers. Linear integrated-circuit audio amplifiers are designed for use as hearing aids, microphone amplifiers, the audio sections of radio and TV receivers, and phonograph preamplifiers for high-fidelity systems.

The small size of an integrated-circuit amplifier has allowed a complete hearing aid, with the amplifier, microphone, earphone, and battery all included in a unit small enough to fit completely within the ear. For the integrated-circuit amplifier, three or four direct-connected stages are used. Because in use a hearing aid is not subjected to wide temperature variations, a relatively simple single-ended circuit with negative feedback for stabilization is possible. However, balanced differential stages cost little more in integrated-circuit form and, in addition to giving better stability, are convenient for driving a class B push-pull output stage to give low battery drain. Hearing-aid circuits are designed to operate from a 1.5-volt battery and have a maximum voltage gain around 4,000 (72 db). Typical silicon-wafer size for a differential-stage type is about 50 mils square.

For microphone and phonograph pickup preamplifiers, a lower gain is sufficient, and more attention is given to tailoring the frequency response, by using combinations of frequency-selective circuits in the feedback network as described earlier (Fig. 6.8). For stereo systems, two identical preamplifiers can be fabricated on a single integrated-circuit wafer, giving good balance under all conditions, with added economy.

For integrated-circuit audio power amplifiers, there are two special considerations. The chips must accommodate two power transistors for the output stage, resulting in a relatively large chip which may adversely affect the yield; and since power transistors operate with a high junction temperature, the whole circuit must be designed to operate at that temperature. There will also be the need for a special package to facilitate heat transfer from the chip.

Wideband and Video Amplifiers.   Wideband and video amplifiers are designed to give a uniform amplification over a very wide band of frequencies, from zero (dc) up to a high frequency, between one and several hundred megahertz, depending upon the application. Video amplifiers for black-and-white TV go up to about 5 Mhz, and for color TV up to between 10 and 15 Mhz. Wideband pulse amplifiers for some radar applications may extend up to 40 Mhz.

The upper frequency response of an amplifier is dependent both on the type of transistor element used and on the frequency characteristics of the transistor load circuits. With simple resistor loads, as the frequency increases, the load impedance will eventually fall, due to the decreasing reactance of the circuit capacitance in parallel with the resistor. At the frequency at which the reactance of the parallel capacitance equals the load resistance, the impedance of the parallel combination has fallen by 30 percent, and so the voltage gain of the stage has also fallen by 30 percent. Low values of load resistance are used (typically a few hundred ohms), and the parallel capacitance is made up of the collector capacitance of the transistor plus the input capacitance of the next stage together with the small stray capacitance of the interconnections. If the load resistance is 800 ohms, and the total capacitance is 4pf, the reactance of 4pf will equal 800 ohms at 50 Mhz, and so the stage gain will have fallen by 30 percent at that frequency.

Integrated-circuit wideband amplifiers can, of course, be designed for any top frequency. A typical value is 40 Mhz, with a voltage gain of 100 or 200. Although single-ended cascade circuits with feedback can be used, the general trend again is to use differential stages with some internal feedback and provision for adding more negative feedback externally. This arrangement allows either single-ended or push-pull working as required. An emitter-follower stage has a low input capacitance and a low output resistance and so is very useful in integrated-circuit wideband amplifiers, especially as a coupling stage between two common-emitter stages. By adding suitable external negative feedback, a wideband amplifier can be converted to a low-pass narrowband amplifier, or by adding tuned circuits at the input and output, it can be used as a tuned amplifier at any frequency within its passband.

High-frequency Linear Integrated Circuits.   Because it is not possible to form inductances greater than a microhenry or so in monolithic form, with tuned high-frequency integrated circuits, it is necessary to add external, discrete tuning elements. However, integration of the rest of the circuit still offers distinct advantages. A typical arrangement of an integrated-circuit tuned high-frequency stage is shown in Fig. 6.10. From the dc viewpoint, the circuit is a differential pair (transistors T_2 and T_3) with transistor T_1 as a constant-current source. The high-frequency signal is fed into transistor T_1, is amplified, and passed to transistor T_2, where it is amplified further to give the output from the tuned circuit in the collector. If a suitable automatic-gain-control (AGC) voltage is applied to transistor T_3, the collector current of T_1 can be diverted from T_2 into T_3 and so reduce the output. This gives a very convenient AGC system. The combination of T_1 and T_2 can be recognized as the *cascode* circuit often used in discrete circuits. All elements within the dashed

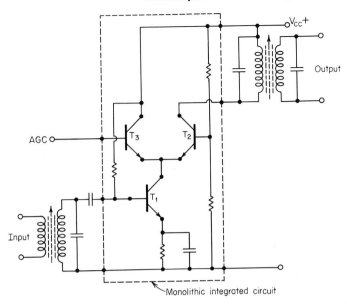

Fig. 6.10. Differential high-frequency linear integrated-circuit amplifier.

enclosure of Fig. 6.10 are fabricated as an integrated circuit. In practice, two or three of these differential circuits are formed in a single integrated-circuit wafer, so that by adding the required external tuned circuits, a complete amplifier can be made using only one integrated circuit. This type of circuit is being applied extensively to intermediate-frequency amplifiers in TV and radio receivers.

Operational Amplifiers. The operational amplifier is the major circuit of the analog computer. It consists essentially of a high-gain amplifier designed to be used with external negative feedback. By using suitably designed negative-feedback networks, the same type of circuit can be arranged to carry out any one of a number of operations required in the analog computer, such as addition, subtraction, multiplication, differentiation, or integration. In each case a standard high-gain "operational amplifier" is used, and the external feedback circuit is designed for the required operation.

The operational amplifier must have a high open-loop voltage gain. For most operations, gains of the order of 30,000 to 40,000 are required, but for some uses, a lower gain of a few thousand times is sufficient. As indicated earlier, the closed-loop operational gain is purely dependent on the external negative feedback. The amplifier is arranged to have a high input impedance and a low output impedance, so that the feedback network does not load the amplifier output and the amplifier input impedance does not affect the feedback voltage.

A typical integrated-circuit operational amplifier consists of two differential-amplifier stages followed by a single-ended output stage. There are two input terminals to the first differential stage. An input to one of these results in an output of the same phase and is called *noninverting*. An input to the other terminal gives an output of the opposite phase and is called *inverting*. When input voltages are

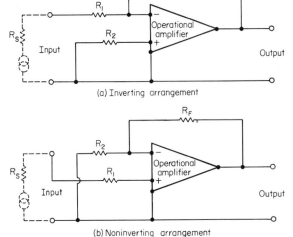

(a) Inverting arrangement

Fig. 6.11. Operational amplifier circuits.

(b) Noninverting arrangement

applied to both terminals, the output is positive or negative, depending on the relative magnitude of the two inputs.

The general method of applying the external feedback is shown in Fig. 6.11. The operational amplifier is represented by a triangle pointing in the direction of the output, the inverting input terminal is marked negative, and the noninverting input terminal positive. When the input signal is connected to the inverting terminal, the connection of the negative feedback is shown in Fig. 6.11a. The output is fed back through a feedback resistor R_F to the inverting input, and a resistor R_1 is included in series between the signal input and the amplifier. R_1 is usually high compared with the signal source resistance R_S, and R_F is high compared with R_1. The fractional feedback β is approximately $(R_1 + R_S)/R_F$, and the closed-loop gain of the system approximates $R_F/(R_1 + R_S)$. The resistance R_2 is made equal to $R_1 + R_S$ to ensure complete symmetry.

With the input signal fed to the noninverting terminal, the feedback is still connected back to the inverting terminal as shown in Fig. 6.11b. The input signal and the negative-feedback signal are compared by the differential circuit to give an output equal to their difference. The closed-loop gain will approximate R_F/R_2.

A practical circuit to give a noninverting amplifier with a precise and stable gain of 10 is shown in Fig. 6.12. An open-loop gain of one or two thousand is sufficient for the amplifier in this case.

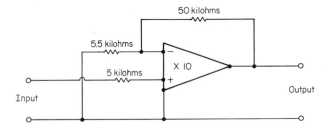

Fig. 6.12. Noninverting gain-of-10 operational amplifier.

The basic method of arranging the negative feedback for the operation of mathematical integration is shown in Fig. 6.13a and, for differentiation, in Fig. 6.13b. In each case, the feedback components are designed to give a predetermined closed-loop gain in addition to carrying out the required mathematical function. The detailed operation of these functional circuits can be read in many textbooks on analog computers.

Differential Comparator Amplifiers. Another category of linear integrated-circuit amplifier is the differential comparator amplifier. Differential stages are used throughout, and the amplifier is intended for use wherever it is required to compare two signals. All types are arranged to take two input signals and usually include three differential stages. Some types have a differential output; others have a power output stage with a single-ended output.

One of the main applications for this type of amplifier is to compare a signal with a reference voltage in order to generate an "error signal" which can be used for control purposes. The system can operate as a simple-level detector, set to give indication when a voltage exceeds a reference level, or the error signal can be used as a feedback signal to result in a constant output or some other required controlled feature.

The differential comparator amplifier forms a basic part of most electronic control systems, and the application of integrated-circuit comparators for this purpose will be discussed in more detail in later lessons.

MOS Linear Circuits. The MOS transistor has the characteristics of high input impedance and high gain, which are both advantageous for linear amplifiers. With the p-channel enhancement MOS transistor discussed in earlier lessons, the gate voltage has the same polarity and can be of a similar value as the drain voltage. This allows a convenient method of obtaining dc bias for linear operation, by connecting the gate to the drain through a high-value resistor as shown in Fig. 6.14a. Since the gate draws no dc current, the full drain voltage appears at the gate. Also, direct connection between stages is readily possible, as the drain potential of one

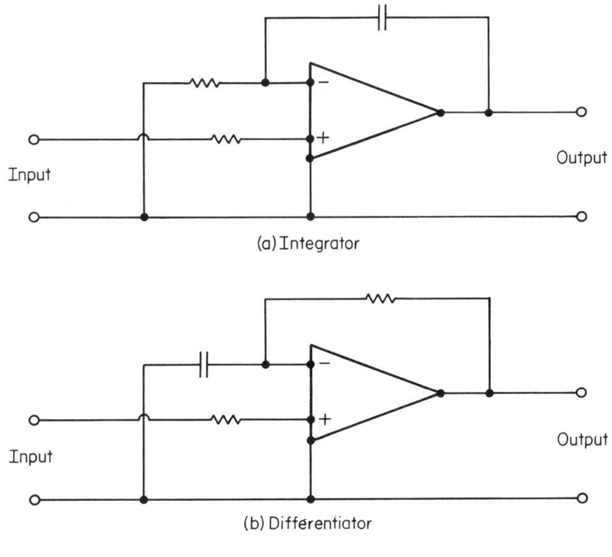

Fig. 6.13. The use of operational amplifiers for analog computing functions.

Input

Output

(a) Integrator

Input

Output

(b) Differentiator

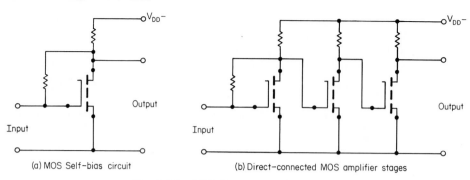

(a) MOS Self-bias circuit   (b) Direct-connected MOS amplifier stages

Fig. 6.14. MOS linear amplifier circuits.

stage will be of the correct order required to bias the gate of the next stage. A direct-coupled three-stage MOS linear amplifier is shown in Fig. 6.14b.

The use of MOS load devices instead of load resistors, as previously described for MOS digital circuits, is also possible with linear circuits. The nonlinear characteristics of the driver and load devices cancel out to give linear operation over a wide range.

MOS linear circuits are very suitable for fabrication in integrated-circuit form, as the MOS structures for the active elements and load resistors are small and simple, and the circuits can be designed with direct coupling to avoid the use of capacitors.

These features combine to suggest a promising future for the MOS transistor in linear integrated circuits. Initially, MOS transistors are being introduced into bipolar integrated circuits to give high input impedance and to combine with bipolar transistors in a Darlington-type circuit to give very high values of transconductance, as shown in Fig. 6.15. Thus the MOS transistor is not only allowing the development of MOS linear integrated circuits which will complement bipolar circuits, but it is giving the possibility of wider scope by combinations of MOS and bipolar devices in the same circuit.

Microwave Integrated Circuits.   Microwave integrated circuits are rapidly becoming a very important group of linear integrated circuits. Because of the large physical size of earlier microwave devices and waveguide components, the dimensions of microwave circuits tended to be relatively large compared with uhf circuits

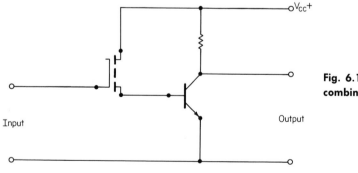

Fig. 6.15. MOS-bipolar combination.

in which lumped components could be used. Semiconductor technology now allows the fabrication of devices with dimensions of the active regions down to the order of 1 micron (1/25 mil), and such devices with expanded contacts together with microstrip lines are opening up the microwave region to integrated circuits, with considerable reduction in size and improvement in efficiency over the previous coaxial and waveguide systems.

Initial considerations were given to the possibility of developing fully monolithic microwave integrated circuits, with the active devices fabricated in a wafer of high-resistivity silicon and interconnected by strip-line elements formed with the silicon as the dielectric. The bottom surface of the silicon wafer would be metallized all over to give a ground plane, and metallized strip leads would be formed on the top surface. Such an arrangement is quite compatible from the dimensional viewpoint. High-resistivity silicon, with resistivity several thousand ohm-cm, has a dielectric constant of 12. With a silicon wafer 10 mils thick, a 50-ohm strip line is obtained with a top conductor approximately 5 mils wide, and an 80-ohm line with a conductor 1.5 mils wide. Such conductors can readily terminate in the expanded contacts of microwave semiconductor devices such as transistors, mixer diodes, and varactor diodes, to give attractive microwave circuits on small silicon wafers.

Unfortunately, there is difficulty in maintaining the resistivity of silicon high enough to prevent excessive dielectric loss. The high-temperature diffusion processes

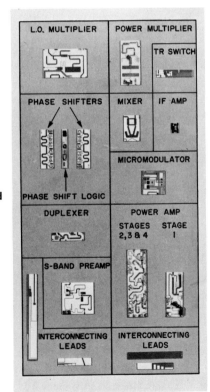

Fig. 6.16. Range of microwave hybrid integrated circuits.

104   Integrated Circuits

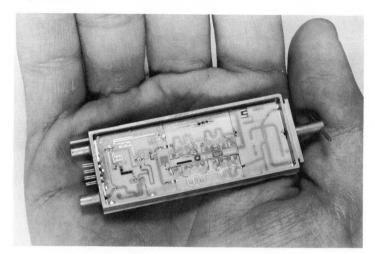

Fig. 6.17.   A microwave integrated-circuit transmit-receive module for an X-band phased-array radar system.

involved in the fabrication of semiconductor devices result in the resistivity of the silicon—initially several thousand ohm-cm—falling below 1,000 ohm-cm. This problem is at present unresolved and as a consequence, an intermediate approach uses ceramic as the dielectric. The strip lines are formed by thin-film metallization on a ceramic substrate, and the semiconductor devices are added as separate "chips."

Using this hybrid approach, a range of microwave integrated circuits has been developed. Figure 6.16 shows a photograph of a number of these circuits. To give an idea of the size, the mixer circuit is 0.5 × 0.32 in. The assembly of these circuits into a complete transmit-receive module for a solid-state phased-array radar system is shown in Fig. 6.17. The module includes two phase-shifting circuits, a power amplifier, two frequency quadruplers, a modulator, a transmit-receive switch, a balanced mixer, and an intermediate-frequency preamplifier—all within the overall dimensions of 2.75 × 1 × 0.75 in. The radiated power is approximately 1 watt peak pulse at 9 Ghz, and the intermediate frequency is 500 Mhz. Several hundred of these modules are used to form a complete phased array for the radar system.

The technology of microwave semiconductor devices is continually advancing the maximum operating frequency. Silicon transistors can be made to operate up to 6 Ghz, and mixer diodes, Shottky barrier diodes, varactor diodes, and Gunn-effect diodes can all be made for operation up to several tens of gigahertz. These devices will be used in an extension of the present hybrid circuits, and there is every possibility that a transition to monolithic techniques using silicon substrates will eventually evolve.

GLOSSARY

closed-loop gain   The overall gain of a system consisting of an amplifier with external negative feedback connected in circuit.

common-mode rejection   The extent to which a differential amplifier stage is immune from signals which are common to both inputs.

differential stage   A symmetrical amplifier stage with two inputs arranged so that with no input signal, a condition of balance exists, and there is no output signal. A signal to either input results in an imbalance, to give an output; but if the same signal is applied simultaneously to both inputs, the balance will still be maintained, and no output signal will result.

inverting input   An input terminal connected to the side of a differential amplifier that results in an output signal of the same phase as the input signal.

linear amplifier   An amplifier in which the swing of the operating point always remains on the linear portion of the forward transfer characteristic so that the output signal is an amplified replica of the input signal.

negative feedback   A system of external feedback from the output to the input of an amplifier such that the feedback signal is in opposite phase to that of the input signal.

noninverting input   An input terminal connected to the side of a differential amplifier that results in an output signal of opposite phase to that of the input signal.

open-loop gain   The gain of an amplifier with no external feedback.

operational amplifier   A high-gain amplifier intended to be used with negative feedback such that the overall characteristics are determined by the feedback network only.

video amplifier   A linear amplifier designed to give linear amplification over a wide range of frequencies from dc up to a few tens of megahertz.

REVIEW

For each of the numbered statements below, select the one of the items lettered *a, b, c,* or *d* that correctly completes the statement.

6.1. For linear amplification, the dc operating point should be set
  a. Near cutoff.
  b. Anywhere on the transfer characteristic.
  c. Midway between cutoff and saturation.
  d. Near saturation.

6.2. One advantage of monolithic integrated circuits for linear applications is that
  a. The temperature of all circuit elements is of the same order.
  b. High values of capacitance can be economically fabricated.
  c. Bias stabilization can readily be effected with close tolerance resistors.
  d. High values of inductance can be obtained.

6.3. When a positive voltage signal is applied to the base of an n-p-n common-emitter amplifier stage
  a. The emitter current decreases.
  b. The collector voltage decreases.
  c. The base current decreases.
  d. The collector current decreases.

6.4. With the Darlington-pair arrangement
  a. The signal current enters both transistor elements in parallel.
  b. The overall current gain is the sum of the current gains of the two transistors.
  c. The overall current gain is the product of the current gain of the two transistors.
  d. A differential output is obtained between the two collectors.

6.5. Differential amplifiers
  a. Have a single input and several outputs.
  b. Cannot be used as single-ended amplifiers.
  c. Are not suitable for fabrication in integrated-circuit form.
  d. Amplify the difference between the two input-signal voltages.

6.6. Negative feedback
  a. Can be used to increase the overall gain.
  b. Cannot be used to vary the frequency response.
  c. Is normally used to give a constant, predicted value of overall gain.
  d. Cannot be used for gain-control purposes.

6.7. Integrated-circuit video amplifiers
  a. Are always relatively narrowband amplifiers.
  b. Only amplify at very high frequencies.
  c. Cannot be converted to narrowband amplifiers.
  d. Give uniform amplification from dc up to typically 40 Mhz.

6.8. With an operational amplifier
  a. If the input signal is connected to the inverting input terminal, the negative feedback must be connected back to the noninverting input.
  b. It is normal to have a high open-loop gain with external negative feedback.
  c. The output impedance is arranged to be high.
  d. The input impedance is arranged to be low.

6.9. Linear MOS integrated circuits
  a. Are not possible.
  b. Cannot use MOS load resistors.
  c. Can be designed to use direct connection between stages.
  d. Are difficult to fabricate.

6.10. Microwave linear integrated circuits
  a. Are designed around waveguide components.
  b. Are presently fabricated using strip-line elements formed by thin-film metallization on ceramic substrates.
  c. Are not suitable for radar.
  d. Have not yet been fabricated.

Standard Catalog Integrated Circuits

INTRODUCTION

So far in this course, the discussions have been centered around basic considerations in order to establish a general understanding of the principles involved in the design, fabrication, and use of integrated circuits. In this lesson, we will review integrated circuits which have become established as standard catalog types, that is, types which are commercially available from stock. For each class of circuit, the general range will be indicated and selected specific types will be described in more detail as typical examples. Although the descriptions mainly refer to types manufactured by the authors' company, they are typical of those manufactured throughout the industry.

It is beyond the scope of this basic course to go into complete detail, particularly with regard to circuit design and integrated-circuit layout. A number of circuit diagrams and chip photographs will be shown, to illustrate the overall circuit complexity possible with integrated circuits and to allow comparison with discrete-device circuits. In showing these detailed circuits, it is not expected that the reader will necessarily be able to appreciate all the details of the circuit design or derive a complete understanding of the internal operation of the circuits. The important thing is to understand the overall function of the circuit as a "black box" with a certain output for defined input conditions.

Integrated circuits are encapsulated in the 14-pin flat pack (type *F*), the dual-in-line plastic package (type *N*), a dual-in-line ceramic package (type *J*), or in a multipin version of the TO-5 transistor package (type *L*). In this lesson, reference to specific circuits will generally be to industrial types packaged in the plastic dual-in-line package (indicated by the suffix *N* after the type number). In general, most types can be obtained in any of the four packages.

STANDARD TYPES OF DIGITAL INTEGRATED CIRCUITS

All the basic digital circuits described in Lesson 5 have been engineered into standard types. The most popular and most widely used circuits are TTL and DTL. In general, RTL and RCTL are rarely used for new system development, and ECL and CTL are used for relatively special applications. Thus in this lesson, emphasis will be on standard catalog TTL and DTL integrated circuits; the other types will not be discussed.

Even with consideration restricted to TTL and DTL, the number of standard circuits is very large, and it will not be possible to review or even list all of the available types. Most standard types are available in two versions, one for military and the other for industrial use.

The range of digital integrated circuits can be divided into two main categories, simple-gate types and combined-gate functional types. The first objective in the development of integrated circuits was to fabricate a complete gate on a single silicon chip and to encapsulate the chip in a suitable package. It soon became clear, however, that several similar gates can be fabricated on a single chip with very little extra cost over that of producing one gate on the chip, and so the next step was to obtain the lowest cost per gate, by forming as many gates as possible on one chip and encapsulating it in one package. The number of gates possible in a single package depends on the total number of connections required for the gate and the number of "pins" on the package. Thus, the simple-gate types consist of variations of the basic gate circuits, repeated as many times as the number of package pins allows.

SIMPLE-GATE TTL TYPES

A group of standard integrated circuits based on the TTL gates described in Lesson 5 is listed in Table 7.1. The circuits of the first four types differ only in the number of input emitters to the logic transistor. The third type, the *dual 4-input positive NAND gate* (SN7420N), is representative of these types. It consists of two separate TTL NAND gates, each with four inputs. The full circuit diagram including both gate circuits is shown in Fig. 7.1a. The internal connections between the circuit chip and the package pins are shown in the logic and package diagram Fig. 7.1b, which gives all the information needed to connect the unit into a system assembly. The supply voltage and ground lines are common to both gates taking up two pins. Each gate requires five pins—four input and one output—making an overall total of twelve pins. The two gates are fabricated on a single silicon chip 45 mils square, shown in Fig. 7.1c.

Table 7.1.   Simple-gate TTL Integrated Circuits

No.	Integrated-circuit type	Industrial type number	Chip size, mils
1	Quadruple two-input NAND gate	SN7400N	50 × 60
2	Triple three-input NAND gate	SN7410N	50 × 60
3	Dual four-input NAND gate	SN7420N	45 × 45
4	Single eight-input NAND gate	SN7430N	40 × 40
5	Dual four-input NAND power gate	SN7440N	50 × 50
6	Triple three-input AND gate	SN74H11	50 × 60
7	Dual four-input AND gate	SN74H21	45 × 45
8	Quadruple two-input NOR gate	SN7402N	50 × 60
9	Dual 2Y two-input AND-OR-INVERT gate	SN7451N	50 × 55

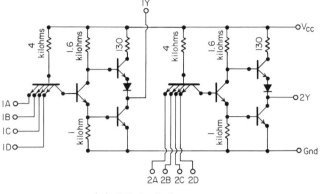

(a) Full-circuit diagram

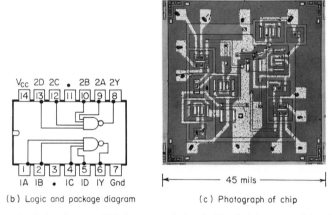

(b) Logic and package diagram   (c) Photograph of chip

Fig. 7.1. A typical simple-gate TTL integrated circuit (dual 4-input positive NAND gate SN7420N).

The operating conditions and performance, which are typical of the whole range of TTL circuits, are as follows:

Supply voltage V_{CC}: 5 volts.

Ambient temperature: -55 to $125°C$ (military types), 0 to $+70°C$ (industrial types).

Power dissipation: 10 mw per gate (50 percent duty cycle).

Propagation delay: 13 nanosec.

Noise margin: 1 volt (0.4 volt specified minimum).

Fan-out: 10.

The dual NAND power gate (SN7440N) consists of two separate 4-input gates, each using the circuit with a power output stage shown earlier in Fig. 5.8c. The AND gates (SN74 H11 and H21) use the circuit shown in Fig. 5.8d. With both of these groups, the power output stage allows a fan-out up to 30.

The quadruple 2-input NOR gate (SN7402N) contains four identical NOR gates which use the circuit shown in Fig. 5.8b. By arranging two inputs to each input transistor of the NOR circuit, an AND function can be introduced in addition to the OR function. This is illustrated in Fig. 7.2*a*. If inputs 1*A* AND 1*B* OR inputs 1*C* AND 1*D* are at logical 1, the gate output (1*Y*) will be logical 0. This type (SN7451N) is called a *dual 2-wide 2-input AND-OR-INVERT gate*. The logic and package diagram is shown in Fig. 7.2*b*, and a photograph of the chip in Fig. 7.2*c*.

Figure 7.3 shows the package diagram for each type listed in Table 7.1. The diagrams show the logic gates and their connections to the package pins.

SIMPLE-GATE DTL TYPES

A family of diode-transistor logic (DTL) digital circuits has also been standardized. Four types representative of DTL simple-gate circuits are listed in Table 7.2.

Typical of these is the *dual 4-input NAND gate* (SN15830N). The full circuit diagram of this type is shown in Fig. 7.4*a*, and the package diagram in Fig. 7.4*b*. The circuit used is a slight modification of the basic DTL circuit shown in Fig. 5.6: the series diode is replaced by a transistor plus a diode as shown in Fig. 7.4*a*. The

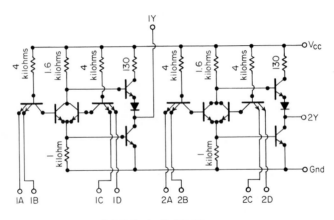

(a) Full-circuit diagram

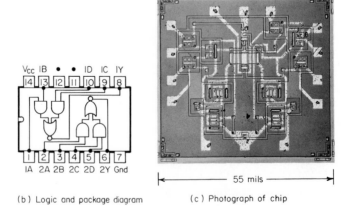

(b) Logic and package diagram

(c) Photograph of chip

Fig. 7.2. Dual AND-OR-INVERT integrated circuit (SN7451N).

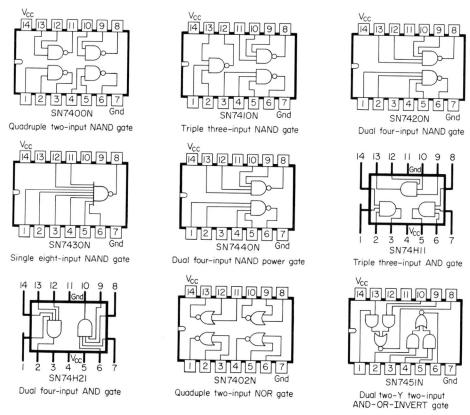

Fig. 7.3. Package diagrams of simple-gate TTL integrated circuits.

gates operate from a 5-volt supply, have a propagation delay of 25 nanosec, a power dissipation of 5 mw, a noise margin of 0.5 volt, and a fan-out of 8. The buffer type (SN15832N) includes a power output stage to give a higher fan-out, up to 25.

COMBINED-GATE CIRCUITS

The simple-gate types described above were used as basic components, and were assembled onto printed circuit boards, the gates being interconnected by the printed-circuit wiring to form complete logic functions. The next stage of develop-

Table 7.2. Simple-gate DTL Integrated Circuits

No.	Integrated-circuit type	Industrial type number
1	Quadruple two-input NAND gate	SN15846N
2	Triple three-input NAND gate	SN15862N
3	Dual four-input NAND gate	SN15830N
4	Dual four-input NAND buffer gate	SN15832N
5	Dual four-input NAND power gate	SN15844N

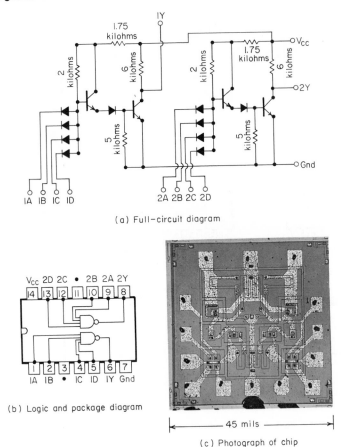

(a) Full-circuit diagram

(b) Logic and package diagram

(c) Photograph of chip

45 mils

Fig. 7.4. Typical DTL simple-gate integrated circuit (dual 4-input NAND gate SN15830N).

ment was to interconnect the gates on the chip by an extension of the metallization pattern. This then gives a self-contained, functional circuit on a single silicon chip. Since a complete circuit needs only input, output, and supply connections, a complex circuit on one silicon chip can be packaged as easily and with no more cost than a single-gate chip, resulting in a lower cost per gate. Complete logic functions including up to 35 interconnected gates have been established as standard types. Table 7.3 lists a group of such standard units based on the TTL system. The number of gates (the complexity) of each type is indicated, as also is the size of the silicon chip for each circuit. The average area of silicon per gate for these circuits is approximately 300 square mils (17 mils square). This clearly indicates the remarkable advance that integrated-circuit technology has achieved; only two or three years ago the smallest planar transistor used a silicon chip 25 mils square (625 square mils).

Four basic functions are included in Table 7.3—flip-flops, registers, binary counters, and adders. A brief description of each function will be given together with more detail of a typical type of each group.

Table 7.3. Combined-gate TTL Integrated-circuit Functions

No.	Integrated-circuit type	Industrial type number	Chip size, mils	No. of equivalent gates
1	J-K flip-flop	SN7470N	55 × 60	8
2	J-K master-slave flip-flop	SN7472N	55 × 60	8
3	Quadruple latch	SN7475N	60 × 120	24
4	Gated full adder	SN7480N	65 × 65	14
5	2-bit full adder	SN7482N	65 × 65	21
6	Decade counter	SN7490N	50 × 115	18
7	8-bit shift register	SN7491AN	55 × 110	35
8	Divide-by-12 counter	SN7492N	50 × 115	17
9	4-bit binary counter	SN7493N	50 × 115	17
10	Dual 4-bit shift register	SN7494N	70 × 110	20
11	5-bit shift register	SN7496N	70 × 140	24
12	5-bit ring counter	SN1286N	70 × 140	30

Flip-flops. The basic flip–flop circuit was mentioned in Lesson 4. It consists of two inverter stages cross-connected as shown in Fig. 7.5a. Each stage is effectively a NOR gate, and the circuit can be represented using logic symbols as shown in Fig. 7.5b. Since the overall operation of the circuit can be defined by the input and output states, the representation of the circuit can be further simplified to a "box" as shown in Fig. 7.5c. The circuit can be triggered into either of two stable states by applying suitable input pulses and will remain in that state indefinitely until triggered again. Thus it will store one bit of information (a logical 1 or a logical 0) as long as is required. The information is fed to the input (the *set* terminal) to trigger the circuit, and when the information is no longer needed, the circuit can be changed back to the original state by applying a signal pulse to the *reset* (or clear) terminal. The state of the circuit is indicated by the output voltage level at either of the two collectors. The two output levels Q and $\overline{Q}$ will always be the complement of each other—when one is at logical 1, the other will be at logical 0, and vice versa. This basic circuit is called an *SR flip-flop* (for Set-Reset).

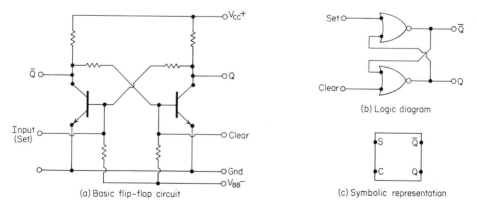

(a) Basic flip–flop circuit

(b) Logic diagram

(c) Symbolic representation

Fig. 7.5. Basic flip-flop details.

In the operation of a system including a flip-flop circuit, it is usual to set and clear the circuit at specific times, at which other circuits are also manipulated. This is carried out by adding "clock" pulses to the inputs so that the circuit only operates if both input and clock signals are present together. A simple two-input AND gate is included in each of the set and reset circuits to accomplish this, and the circuit is then called a *clocked SR flip-flop*.

A widely used flip-flop circuit with additional logic systems in the input circuits is called a *J-K flip-flop*. This has two input terminals designated *J* and *K*. The circuit can be set by a signal pulse applied to the *J* input and cleared by a pulse applied to the *K* input. In addition, a signal applied simultaneously to both *J* and *K* inputs will trigger the circuit from either state to the other. This is called the trigger input.

A typical standard circuit is the second in Table 7.3, the *J-K master-slave flip-flop* SN7472N. In the "master-slave" arrangement, the inputs to a master flip-flop stage are controlled by the clock pulses as usual, and the information is then transferred to a second flip-flop circuit (the "slave" circuit), the transfer also being controlled by the clock pulse. Output is taken from the slave stage. The full circuit diagram of the type SN7472N is shown in Fig. 7.6a, the logic diagram in Fig. 7.6b, and the package diagram in Fig. 7.6c. In addition to the clock logic circuits, there are two AND gates in the *J* and *K* inputs, the total circuit comprising the equivalent of eight gates as shown in Fig. 7.6b. The state of the output circuits can be set independently of both the clock and the *J-K* inputs by signals fed to the *preset* and *clear* terminals. Operating conditions and logic voltage levels are the same as the general TTL range. Total power consumption of the circuit is 40 mw ($V_{CC} = 5$ volts), and total propagation delay through the circuit is 30 nanosec.

Flip-flop circuits form the basic logic element of binary information stores, called registers, and also for binary-counting circuits. These will be discussed in turn.

Registers.   The most common type of register is the serial or *shift register*. In this system, a number of flip-flop circuits are connected in series, the output terminals of one being connected to the input terminals of the next. Clock pulses are fed to all the flip-flop circuits. When a clock pulse appears, information at the set input is transferred into the first flip-flop where it is stored. When the next clock pulse appears, this information stored in the first circuit is transferred to the second circuit, and a new input enters the first circuit. Every time a clock pulse appears, information is shifted from each flip-flop circuit into the next succeeding one, and a new bit of information enters the first circuit. In this way, several bits of information can be stored, 1 bit per flip-flop stage. The information can be stored for a predetermined time—milliseconds, minutes, or hours—and on command can be presented at the output for further use.

A typical standard integrated circuit of this type is item 7 in Table 7.3, the *8-bit shift register* SN7491AN. An SR flip-flop, somewhat less complex than the *J-K* flip-flop described above, is used for each bit. The logic connections are shown in the package diagram Fig. 7.7a. Input data are fed through an input gate and an inverter stage to give complementary signals to the *S* and *R* inputs of the first flip-flop. The clock pulses are fed to an inverting driver stage which drives the common clock line. Typical input and output waveforms are shown in Fig. 7.7b.

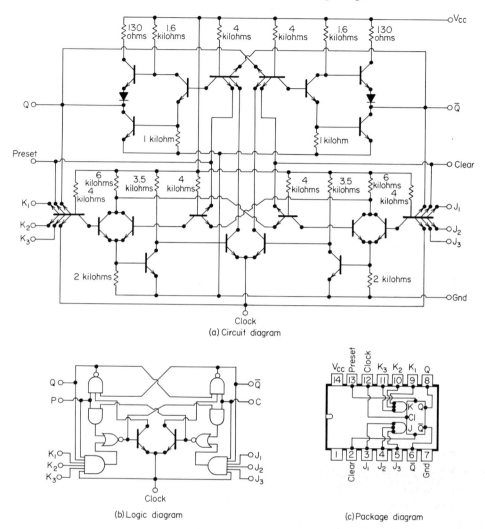

(a) Circuit diagram

(b) Logic diagram

(c) Package diagram

Fig. 7.6. TTL J-K master-slave flip-flop (SN7472N).

An input signal appears at the output eight clock pulses later; the width of the output signal being somewhat greater, extending to the start of the next clock pulse. Each flip-flop includes 4 gate circuits, and so the total integrated circuit includes the equivalent of 35 gates. The silicon chip is 55 × 110 mils and is shown in Fig. 7.7c.

TTL logic is used, the total power dissipation is typically 175 mw, clock-pulse repetition rate can be up to a maximum of 18 Mhz, and a fan-out of 10 is possible at the output.

Binary Counters. Flip-flop circuits can be interconnected to give binary counting as shown in Fig. 7.8a. When the first input pulse occurs, the first flip-flop (FFA) changes state to give a logical 1 at the output A. The second input pulse will trigger the circuit back to give a logical 0, the third a 1, the fourth a 0, and so on. The input circuits of the flip-flops are all arranged so that they trigger to the opposite

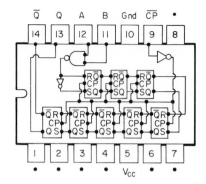

(a) Logic and package diagram

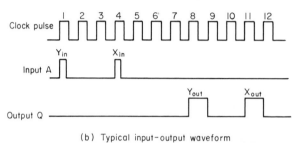

(b) Typical input–output waveform

Fig. 7.7. 8-bit shift register (SN7491AN).

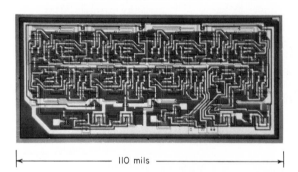

(c) Photograph of chip

state at the trailing edge of the input pulse. Thus, when FFA changes from 1 to 0, the second flip-flop (FFB) will trigger to 1 and stay in that state until FFA changes from 1 to 0 again, two input pulses later; then FFB triggers back to 0. Similarly, the output from FFB will trigger FFC, and the output from FFC will trigger FFD. The waveforms at the outputs of the flip-flops are shown in Fig. 7.8b. It will be seen that the binary count of the input pulses is given at any time by the D-C-B-A sequence of the binary states at the outputs. For example, immediately after the seventh input pulse, we have 0111, which is the binary equivalent of 7. After the twelfth input pulse, we have 1100, and so on.

A standard integrated circuit of this type is the *4-bit binary counter* SN7493N.

The functional logic diagram is shown in Fig. 7.8c. Four *J-K* flip-flops are used. The first is arranged to be separate and gives a count of 2, and the other three are interconnected to give a count of 8. By externally connecting the output of the first to the input of the second, a count of 16 can be obtained. A gate reset line is included; a signal applied to this simultaneously returns all flip-flops to logical 0. The circuit is fully compatible with other TTL circuits. Average power dissipation is 155 mw with a supply voltage of 5 volts.

Adding Circuits. The general principles of binary addition were described in Lesson 4. Complete adding circuits in a single package have been established as standard integrated circuits. Typical of these is the *2-bit binary full adder* SN7482N. The system used in this circuit is somewhat different from the full adder described in Lesson 4. With that system, for a 2-bit full adder with carry-in, a total of 26 gates are required. By using the arrangement of the type SN7482N, the number

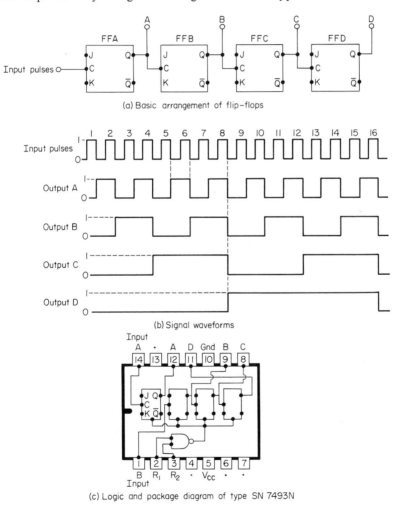

(a) Basic arrangement of flip-flops

(b) Signal waveforms

(c) Logic and package diagram of type SN 7493N

Fig. 7.8. Integrated-circuit binary counter.

of gates is reduced to 21. The logic diagram is shown in Fig. 7.9. It is suggested that the reader work through the operation of this arrangement by defining the logic conditions at the inputs and outputs of the various gates, starting with the inputs A, B, and C_{in}.

The 21 gates are all formed on a single silicon chip 65 mils square. In another standard type, the SN7483N, two of these chips are mounted in a single package to give a 4-bit binary full adder.

Combined-gate DTL Circuits.   A range of combined-gate circuits using diode-transistor logic (DTL) has also been established as standard types. The functions are generally similar to the above TTL functions and so will not be discussed further.

STANDARD MOS DIGITAL INTEGRATED CIRCUITS

The use of the MOS structure for both the active switching element and the load resistor, coupled with the very small area it occupies, gives the possibility of relatively complex digital logic circuits on quite small silicon chips. As a result, simple-gate MOS types have not generally been established as standard types—the industry has gone directly to functional circuits such as full shift registers, counters, and adders. One or two simple-gate types have been made available—intended not for use in systems, but for use in "breadboard" development of new system design.

A list of standard MOS digital integrated circuits typical of those commercially available is given in Table 7.4. Multistage shift registers are ideal for MOS fabrication; the stages can be formed in line and interconnected by simple metallization stripes. A very complex shift register still only needs input, output, and clock and supply connections and so can be encapsulated in a simple package.

MOS Shift Registers.   With MOS circuits, two types of shift registers have been established, static and dynamic. The static type uses several flip-flop circuits in an arrangement similar to that of the bipolar registers described earlier. A typical standard MOS shift register is the *dual 50-bit static shift register* TMS7C3002LA. This consists of two separate 50-bit shift registers with independent input and output

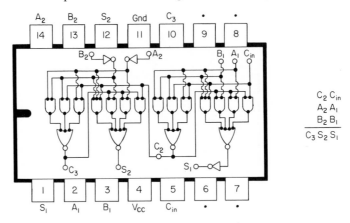

Fig. 7.9.   2-bit binary full adder (SN7482N).

Table 7.4. Typical Standard MOS Digital Integrated Circuits

No.	MOS integrated circuit	Type number	Chip size, mils	Equivalent gates per chip
1	Dual 16-bit static shift register	TMS1B3016LA	56 × 36	66
2	Dual 25-bit static shift register	TMS7B3000LA	70 × 70	102
3	Dual 32-bit static shift register	TMS7B3001LA	81 × 69	130
4	Dual 50-bit static shift register	TMS7C3002LA	86 × 80	202
5	Dual 100-bit static shift register	TMS7C3003LA	140 × 90	402
6	Dual 25-bit dynamic shift register	TMS7D3300LA	75 × 48	52
7	4-bit up-down counter	TMS7A3801FB	81 × 64	42
8	Dual full adder	TMS1A1700AA	55 × 42	16
9	Dual three-input NOR gate	TMS1A1702LA	60 × 35	2

terminals and common clock, power supply, and ground lines.   MOS p-channel enhancement transistors are used; the full dual circuit containing the equivalent of 202 gates (two gates per bit) is formed on a silicon chip 80 × 86 mils.   A photograph of the chip is shown in Fig. 7.10.   Cross-coupled flip-flop circuits are used for each stage, allowing information to be stored indefinitely between clock pulses. The circuit of a single stage is shown in Fig. 7.11a.   Three different clock pulses are used to ensure correct triggering sequence of the circuit.   One of the clock pulses

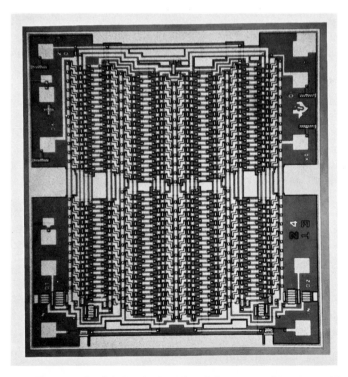

Fig. 7.10. Photograph of MOS dual 50-bit shift-register chip (86 × 80 mils).

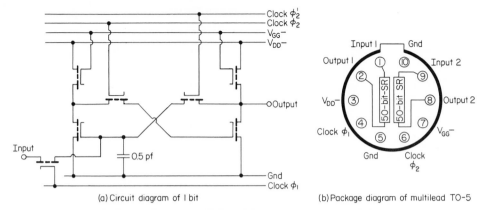

(a) Circuit diagram of I bit

(b) Package diagram of multilead TO-5

Fig. 7.11.   MOS dual 50-bit shift-register TMS7C3002LA.

is usually generated in the circuit.   It will be seen that the circuit is formed completely with MOS elements.

The package diagram is shown in Fig. 7.11b.   A multipin circular TO-5 package is used for this type, but it would appear that the dual-in-line plastic package which has become the popular standard for bipolar industrial circuits will also become a standard package for MOS circuits.

Typical operating voltages are drain voltage, -14 volts; gate bias voltage, -28 volts; clock pulse level, logical 1, -28 volts; logical 0, -2 volts; signal input level, logical 1, -10 volts; logical 0, -2 volts.   Maximum clock-pulse repetition rate is 1 Mhz, typical power dissipation is 210 mw, and the operating temperature range is 0 to 85°C.

In the MOS *dynamic shift register* the information is stored by charging up the small gate capacitance of an MOS structure.   Because the gate is insulated by a layer of silicon oxide, the charge will remain for an appreciable length of time before it leaks away.   A time of a few milliseconds is sufficient for many storage applications, but this means that there will be a lower frequency limit below which the circuit will not operate.   A four-line clocking system is used, and this allows a faster operating speed.   A typical dynamic shift register is the TMS7D3300LA, listed in Table 7.4.

Other MOS Standard Types.   The range of MOS digital integrated circuits is being broadened to include counters, adders, and other logic functions.   Items 8 and 9 on Table 7.4 are typical of those already available.   As mentioned earlier, simple-gate types such as the TMS1A1702LA, a dual 3-input NOR gate, are available for use during complex circuit development.

STANDARD LINEAR INTEGRATED CIRCUITS

The basic aspects of linear integrated circuits were discussed in Lesson 6, and the various categories of circuits that have evolved were described.   Table 7.5 lists a range of standard linear integrated circuits intended for industrial applications.   Several of these types will be described as typical of their class.   As with the digital types, the

Table 7.5. Standard Linear Integrated Circuits

No.	Linear integrated-circuit type	Type number
1	General-purpose operational amplifier	SN72702N
2	High-gain operational amplifier	SN72709N
3	Differential comparator	SN72710N
4	Dual differential comparator	SN72711N
5	General-purpose differential amplifier	SN723
6	General-purpose operational amplifier	SN724
7	High-gain differential amplifier	SN725
8	General-purpose differential amplifier	SN726
9	Wideband video amplifier	SN7510
10	Wideband differential amplifier	SN7511
11	Sense amplifiers	SN7520/25N

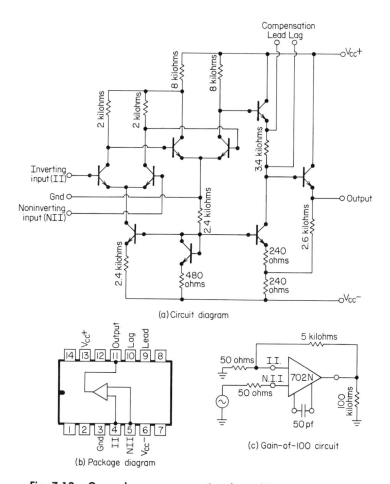

(a) Circuit diagram

(b) Package diagram

(c) Gain-of-100 circuit

Fig. 7.12. General-purpose operational amplifier (SN72702N).

industrial linear types are characterized for operation over a temperature range of 0 to 70°C, and the most popular package is the dual-in-line type N plastic package.

Operational Amplifier. The type SN72702N is typical of medium-gain operational amplifiers. The full circuit diagram is shown in Fig. 7.12a, and the package diagram in Fig. 7.12b. The amplifier has differential inverting and non-inverting high-impedance inputs and a single-ended low-impedance output. Two differential amplifier stages are used, followed by a driver stage and an emitter-follower output stage. Provision is made to allow external compensation components to be added to ensure stable operation under various feedback conditions. The circuit is suitable for any application requiring the transfer or generation of linear or nonlinear functions up to a frequency of 30 Mhz.

Operating with a positive supply voltage of 12 volts and a negative line of 6 volts, typical open-loop gain is 2,600 (about 67 db), and the common-mode-rejection ratio is 10,000 to 1 (80 db). The dc drift referred to the input is 7 microvolts per °C. Input resistance is 20,000 ohms, and output resistance 200 ohms. Typical power dissipation is 80 mw.

The use of the amplifier to give a gain-of-100 operational amplifier is illustrated in Fig. 7.12c.

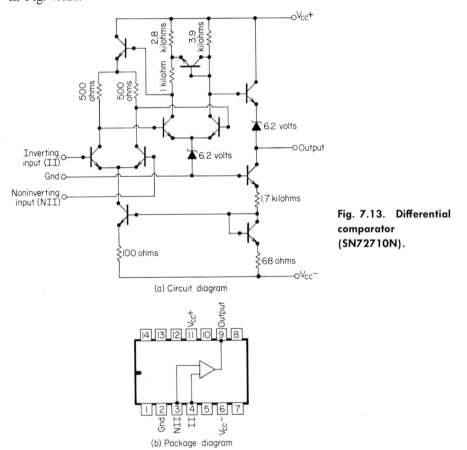

(a) Circuit diagram

(b) Package diagram

Fig. 7.13. Differential comparator (SN72710N).

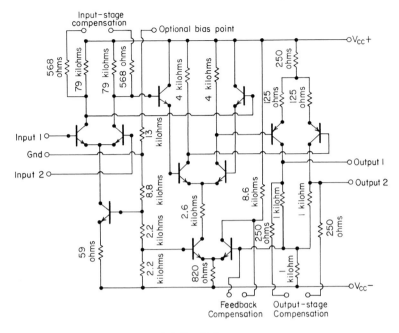

(a) Circuit diagram

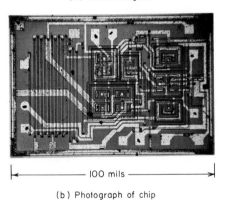

├─────── 100 mils ───────┤

(b) Photograph of chip

Fig. 7.14. High-gain general-purpose differential amplifier (SN725N).

Differential Comparator. Typical of standard differential voltage comparator amplifiers is the integrated-circuit type SN72710N. The circuit includes two differential stages with a low-impedance output stage as shown in Fig. 7.13a. The package diagram is shown in Fig. 7.13b. Typical voltage gain is 1,200 and the dc drift referred to the input is 7 microvolts per °C. The circuit is particularly attractive for voltage comparison where high response speed is required.

High-gain Differential Amplifier. A comprehensive range of high-performance general-purpose differential amplifiers has been established. Typical of this group is the SN725N which has an open-loop gain of 20,000 (86 db) on each side. The full circuit diagram is shown in Fig. 7.14a. Three differential stages are used, the

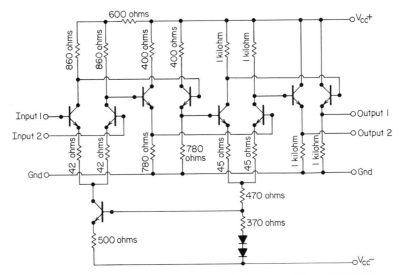

Fig. 7.15. Circuit diagram for wideband video amplifier (SN7510N).

center stage consisting of a high-gain Darlington-pair arrangement. An amplifier stage is included in the common-mode feedback circuit to give a very high common-mode rejection. The output stage uses p-n-p transistors to facilitate direct coupling. The chip size is 90 × 60 mils and a photograph of the chip is shown in Fig. 7.14*b*.

Provision is made for connecting external compensating components at three points. The frequency characteristic can be adjusted by external networks connected to the input stage, the output impedance can be adjusted by connecting external components to the output stage and stability can be ensured by connecting suitable components in the common-mode feedback circuit. For stable open-loop operation, an external 250pf capacitor must be connected to the common-mode feedback circuit.

Operating with positive and negative supplies of 12 volts, the power dissipation is around 100 mw, common-mode rejection ratio is about 30,000 to 1 (90 db), and the dc drift referred to the input is 5 microvolts per °C.

Wideband Video Amplifier.   A number of single-ended wideband amplifiers have been established as standard types, but the trend is toward the differential arrangement, which generally gives better performance and wider application. A typical standard differential video amplifier is the SN7510N (Fig. 7.15). It consists of four stages, an input differential stage with common-mode feedback, coupled by emitter followers to another differential stage and emitter-follower output circuits. It has an open-loop gain of 100 and a flat frequency response from dc to 40 Mhz. The input resistance is 6,000 ohms, the input capacitance 7 pf, and the output impedance 35 ohms. Internal negative feedback is incorporated, resulting in good open-loop stability. With external feedback, useful gain up to 100 Mhz can be obtained. By connecting tuned circuits at the input and output, a narrowband amplifier can be obtained at any frequency within the passband of the circuit.

The SN7510N circuit operates from positive and negative supply lines of 6 volts and has a power dissipation of 165 mw.

Sense Amplifiers.   A group of standard amplifiers called sense amplifiers combine linear and digital functions.   A sense amplifier is designed to accept low-level pulse signals—for example, from a memory system—to amplify them, and then to transform them into logical levels compatible with logic circuits used for subsequent logic operations.   A typical range of standard sense amplifiers is the series SN7520N.   There are three basic parts of a sense amplifier, a differential input amplifier, a differential voltage-level detector, and a pulse-level-adjusting output circuit.   The input signal is amplified and applied to the level detector, where it is compared with a reference voltage.   If it exceeds the reference voltage, it passes to the output circuit, where it is adjusted to give a logical 1 level compatible with TTL logic (2.4 volts minimum).   If the amplified signal is less than the reference voltage, it is rejected, and the output circuit gives a logical 0 output (less than 0.4 volt).

COMMENT

In this lesson we have looked at only a few of the hundreds of standard commercial integrated circuits.   The choice of types discussed was made so as to give the reader a general appreciation of what is being accomplished with integrated circuits.   The details of specific circuits given in this lesson will have helped to indicate the types of integrated circuits commercially available.   The emphasis should now be on what function the integrated circuit performs, rather than on the details of how it does it.   It cannot be overemphasized that anyone proposing to use integrated circuits should consult the manufacturer for advice and full details of his particular circuits.

GLOSSARY

binary counter   A system of flip-flop circuits and gates with a single input designed to count the number of times the input changes state in a given direction.

clocked flip-flop   A flip-flop circuit which is set and reset at specific times by adding clock pulses to the inputs so that the circuit is only triggered if both trigger and clock pulses are present simultaneously.

combined-gate integrated circuit   An integrated circuit consisting of a number of gate circuits combined by interconnection on the chip to give a more complex circuit.

dynamic shift register   A shift register in which the information is stored by charging a small capacitor.   Since the charge will leak away at a rate depending on the circuit, the information can only be stored for a relatively short time.

flip-flop circuit   A circuit having two stable states, arranged so that it can be triggered from one state to the other by an input trigger pulse.   In the absence of a trigger pulse, the circuit will stay permanently in the state which exists.

J-K flip-flop   A flip-flop circuit with two input terminals designated J and K.   The circuit can be "set" by a pulse applied to the J input and "reset" by a pulse applied to the K input.   In addition, a pulse applied simultaneously to both J and K inputs will trigger the circuit from the existing state to the other.

reset terminal   The input of a flip-flop circuit to which a trigger pulse is applied to trigger the circuit back from the second state to the original state.   It is alternatively called the *clear* terminal.

sense amplifier   An amplifier system designed to accept and amplify small signals such as from a transducer and then transform them into discrete logical voltage levels compatible with subsequent logic-circuit operation.

set terminal   The input of a flip-flop circuit to which the trigger pulse is applied to trigger the circuit from the first state to the second state.

shift register   A system designed to accept information in the form of a series of electronic pulses, to store it for a period of time, and on command to feed the same information from the output.

simple-gate integrated circuit   An integrated circuit consisting of one or more complete but separate gate circuits, the input and output of each gate being connected to separate pins on the package.

static shift register   A shift register consisting of a number of series-connected flip-flop circuits capable of storing information for an indefinite period of time.

REVIEW

For each of the numbered statements below, select the one of the items lettered *a, b, c,* or *d* that correctly completes the statement.

7.1. The most popular and most widely used logic integrated circuits are
  a. Simple-gate types.
  b. TTL and DTL.
  c. RTL and RCTL.
  d. ECL and CTL.

7.2. A dual 4-input gate integrated circuit consists of
  a. A single gate with eight inputs.
  b. Four gates each with two inputs.
  c. Two separate gates each with four inputs
  d. A single gate with two groups of four inputs.

7.3. In combined-gate monolithic integrated circuits
  a. Several separate chips, each with one gate, are assembled into one package.
  b. The total number of gates is generally below ten.
  c. Only TTL logic is used.
  d. A number of gate circuits are interconnected on a single chip to give a complete logic function.

7.4. A flip-flop circuit
  a. "Flips" from one state to the other when a trigger pulse is applied and then "flops" back after a predetermined time.
  b. Is usually set and cleared at specific times by adding clock pulses to the inputs.
  c. Is basically two cross-connected AND gates.
  d. Cannot be used for binary counting.

7.5. In a serial or shift register
  a. A bit of information is fed into the input and shifted to the last flip-flop before the second bit is fed in.
  b. Several bits of information are fed in simultaneously to several inputs.
  c. It is possible to store two bits of information per flip-flop stage.
  d. Information pulses are always fed into the first stage; with each clock pulse, the information is shifted from each stage to the next succeeding one, and a new bit of information is fed into the first stage.

7.6. A binary counter
   a. Has a single input and counts the number of times the input changes state in a given direction.
   b. Has several inputs and is arranged to count the number of times the output changes state.
   c. Is normally made of a group of separate-gate circuits.
   d. Is the same as a shift register.

7.7. MOS integrated-circuit shift registers
   a. Are being made with 100 bits on a chip 85 mils square.
   b. Can only be of the static type.
   c. Usually incorporate diffused load resistors.
   d. Are generally of the dynamic type to store information indefinitely.

7.8. A typical medium-gain integrated-circuit operational amplifier
   a. Has only one input terminal.
   b. Is only suitable for use at low frequencies.
   c. Has an open-loop gain about 2,500.
   d. Has an open-loop gain about 100.

7.9. A typical high-gain integrated-circuit differential amplifier
   a. Has a power dissipation about 100 microwatts.
   b. Has no provision for connecting external compensating components.
   c. Consists of one Darlington-pair stage.
   d. Has three differential-amplifier stages with an open-loop gain around 20,000.

7.10. An integrated-circuit sense amplifier
   a. Is a memory system.
   b. Consists of three separate linear amplifiers.
   c. Consists of a linear amplifier, a voltage-level detector, and a logical pulse-forming circuit.
   d. Cannot practically be made as a standard product.

Integrated Electronic Components

INTRODUCTION

During the manufacture of integrated circuits, several hundred identical circuits are formed on a single slice of silicon, and the slice is then cut up into separate circuits, which are assembled, packaged, and tested individually. In many digital applications, such as computers, large numbers of identical logic gate circuits are used in a repetitive manner. To fabricate such a functional logic system, the individual integrated circuits are reassembled onto boards and interconnected to give the required system. The obvious question is: "Why not interconnect the required number of logic circuits on the silicon slice instead of cutting it up?" Then the complete logic system would be formed on one piece of silicon which could be packaged as a single unit.

Actually, this approach has already started. In the previous lesson, it was indicated how the complexity of digital integrated circuits, that is, the number of gate circuits formed on a single chip of silicon and encapsulated into a single package, has increased from about three gates for the simple-gate types to about twenty gates for the combined-gate types. Several reasons can be listed to explain why it is sensible to include more circuits on a single silicon chip:

1. The package and assembly represent more than half the cost of a simple-gate integrated circuit. If more gate circuits are included on the chip, resulting in more gates per package, the package cost will be shared between several gates, giving a lower total cost per gate.

2. A considerable fraction of the area of an integrated-circuit chip is taken up by the bonding pads for terminating the connecting wires from the package leads to the chip. To include more gates on the chip, it is only necessary to increase the chip area for the additional circuits—the bonding-pad area remains the same. Thus the effective chip area per gate is reduced, giving a lower cost per gate.

3. With more circuits contained within the integrated-circuit package, there will be a corresponding reduction of printed-circuit-board assembly and system housing, both contributing to a reduction in system cost.

4. The overall size and weight of the complete equipment will be smaller,

allowing lower freight and installation costs and a lower operating cost due to floor rental, etc.

5. The greatest cause of failure of electronic equipment is undoubtedly due to faulty interconnections between components, and so the reliability of electronic equipment is an inverse function of the number of electrical connections to the pins of the components. The metallized interconnections on the surface of a silicon integrated circuit are extremely reliable, being formed in one closely controlled process on many circuits at the same time. Thus, the greater the complexity on the silicon chip and the smaller the number of external connections, the higher will be the system reliability.

Thus, there are strong motivations to increase the complexity of integrated circuits in order to obtain lower system cost and higher system reliability. As mentioned above, this approach is particularly suitable for digital integrated circuits because of the repeated use of identical logic gates. The considerations do not apply so much with linear integrated circuits, since there is little repetition of circuits in linear systems. Nevertheless, there is continual progress in the design of all integrated circuits to include more circuits on a chip, interconnected by the metallization pattern to form complete functional units. These complex circuits are called *integrated electronic components* (IEC). Since this is a new class of component, it will do well to define it, particularly with relation to the other semiconductor products.

DEFINITIONS

The full range of semiconductor products can be classified into three categories: discrete devices, integrated circuits, and integrated electronic components. The definitions given here are not necessarily accepted throughout the industry.

Discrete Device. This is an individual semiconductor element encapsulated in a suitable package. The element will normally consist of a single transistor or diode but may consist of a dual structure, for example, two transistors formed in the same wafer for thermal matching.

Integrated Circuit. A monolithic integrated circuit can be defined as an electronic circuit which has been fabricated as an inseparable assembly of semiconductor elements in a chip of silicon with a circuit content up to 10 equivalent gates. The circuit may consist of several separate logic gates or simple circuit functions, such as a flip-flop circuit or a differential amplifier.

Integrated Electronic Component. An integrated electronic component is the result of interconnecting a number of integrated circuits on a single chip of silicon to give a complete electronic function with a circuit content greater than 10 equivalent gates.

By this last definition, most of the combined-gate digital circuits described in the previous lesson and listed in Table 7.3 come into the category of integrated electronic components.

In this lesson, we will discuss the various aspects involved in producing IECs, such as the relation between chip area and yield, the relation between complexity and cost per gate, the optimum complexity for lowest cost per gate, and different methods of fabricating high-complexity units both standard and custom.

CHIP AREA VERSUS COMPLEXITY AND YIELD

To include more circuits on a single chip of silicon generally means that the area of the chip must be increased. As was stated in Lesson 1, the yield of good chips from a silicon slice decreases as the area of the chip is increased. A typical relation between the area of an integrated-circuit chip and the yield of good chips on a slice is shown in Fig. 8.1. Thus, to obtain the highest yield of a particular circuit, it should be designed to occupy as small an area as possible. This appears to be in contradiction to the proposal to include more circuits on a chip, which must require some increase in area with an adverse effect on the yield. Obviously the question needs careful evaluation. The area of a chip does not increase linearly with complexity. The bonding pads for making contact to the chip plus the spacing between chips on the slice can occupy up to 35 percent of the area of a simple-gate chip. In increasing the complexity of a chip, only the circuit area is increased —the bonding-pad area remains the same. Thus, if the complexity is increased three times, the total chip area is only increased just over twice, and the reduction of yield due to the increase in complexity is not as great as might appear at first sight. Nevertheless, the reduction in yield must result in corresponding increase in the chip cost, and so it is necessary to evaluate the effect of this on the overall cost per gate.

EFFECT OF INCREASING COMPLEXITY ON COST PER GATE

When considering the cost of integrated circuits and integrated electronic components, it is convenient to use the cost per gate as the basis of comparison. There are two main constituent parts to the overall cost of a unit: first, the cost of fabricating

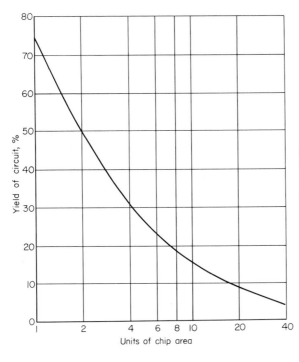

Fig. 8.1. Typical relation between the fabrication yield and chip area of an integrated circuit.

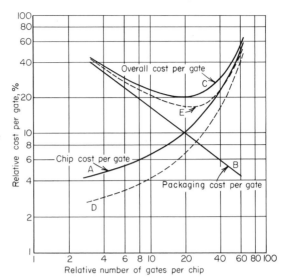

Fig. 8.2. Relation between cost per gate and the chip complexity of integrated circuits.

a good chip and, second, the cost of packaging the chip (this includes the cost of the package, the cost of assembling the chip and sealing the package, and the cost of testing the unit). As the complexity of a circuit is increased, the cost per gate of the chip increases due to the decrease in yield as discussed above. This is shown by curve A of Fig. 8.2. Against this, the overall cost of packaging remains approximately constant, and as it is now shared amongst a larger number of gates, the packaging cost per gate reduces as shown by curve B. The overall cost per gate is the sum of the two, and because of the opposing trends of the two components, it passes through a minimum value. This is very important; it indicates that there is an optimum complexity of circuit that should be encapsulated in a given package to obtain a minimum overall cost per gate.

The effect of increasing the chip yield can be seen in Fig. 8.2. Increased chip yield means reduced chip cost as shown by the dashed line D. The new overall cost is the dashed line E. It will be seen not only that the minimum cost per gate has fallen, but that it has also moved to a higher complexity. This emphasizes the importance of designing the layout of complex integrated circuits to occupy as small an area as possible in order to keep the yield as high as possible. The shape of the overall cost curve around the minimum value is fairly flat, and this allows some departure from the optimum complexity in practical designs with only slight increase in cost.

INTEGRATED-CIRCUIT COMPLEXITY VERSUS SYSTEM COST

The effect on system cost of using optimum complexity integrated circuits is shown in Fig. 8.3. The solid curve is the overall cost per gate versus complexity taken from Fig. 8.2. Using a simple-gate integrated circuit incorporating, say, three gates (point A), the cost per gate will be at point 1. For an optimum-complexity circuit (point B), the cost per gate will be lower, at point 2. If a sufficient number of three-gate circuits are assembled on a printed circuit board to give the same com-

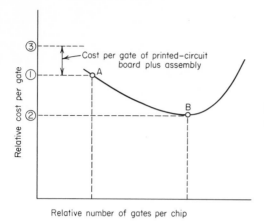

Fig. 8.3. **Effect of integrated-circuit complexity on system cost per gate.**

plexity as the optimum, there will be an additional cost (that of the printed circuit board and assembling the units on the board) giving an overall cost at point 3. Thus the system cost saving by using the optimum-complexity circuit is the difference between points 3 and 2.

INTEGRATED ELECTRONIC COMPONENTS

We see then, that integrated electronic components are a logical extension of integrated circuits into higher levels of complexity. In order to qualify the levels of complexity and the technologies involved in fabricating IECs, the terms *medium-scale integration* (MSI) and *large-scale integration* (LSI) have been introduced. Unfortunately these terms were initially used by different companies to mean different things, resulting in some confusion. Although the terms have not been officially defined, there is now growing acceptance that they relate primarily to the level of complexity, and following the definition of an IEC quoted earlier, the terms MSI and LSI can be defined as follows:

> *Medium-scale Integration* (MSI) refers to the group of technologies used in the fabrication of IECs with circuit complexities in the range 10 to 100 gates.
> *Large-scale Integration* (LSI) refers to the group of technologies used in the fabrication of IECs with circuit complexities above the level of 100 gates.

Since MSI and LSI each relate to several different technologies, it is necessary to qualify them when discussing any particular IEC. For example, the combined-gate circuits in Table 7.3 can be described as using TTL-MSI technology, and those in Table 7.4, MOS-MSI.

Although precise boundaries are included in the definitions, there is no sudden change. Technologies established for simple-gate integrated circuits extend well into the MSI region before additional techniques are necessary, and similarly, MSI techniques spread into the LSI range. As technologies are improved and refined, the need for redefinition may arise.

MEDIUM-SCALE INTEGRATION

Standard Fixed-interconnect MSI.   As indicated above, the first group of circuits falling into the class of MSI is the range of combined-gate circuits such as those listed in Table 7.3.   These range in complexity from 10 to about 35 gates and are fabricated using the same basic technology as that used for the simple-gate integrated circuits.   The interconnections between the individual elements of a gate together with those connecting the gates into the functional circuit are all made by a fixed pattern of metallization on top of the "planar" oxide covering the surface of the slice.   These can all be referred to as standard catalog, TTL-MSI types.

As the complexity of an integrated circuit is increased, one limitation is with the routing of the interconnections, which statistically becomes increasingly difficult without numerous "crossovers."   To reduce the number of crossovers, the elements can be spaced out, but this results in larger chips with lower yield.   To overcome the difficulty, techniques for using two levels of interconnection have been developed. Some of the interconnections are made in the normal way on top of the original planar oxide—this is the first level of metallization.   Then a second layer of silicon oxide is deposited all over the slice, and the remaining interconnections are made on top of this second oxide by a second level of metallization.   This process allows the interconnections on the top layer to cross over those on the first layer and gives the possibility of a smaller chip.   Figure 8.4 illustrates the principle of two-level interconnection, and an example of the reduction in chip size that two-level interconnection allows is shown in Fig. 8.5.   This two-level system is one of the specific technologies that have been developed for MSI.   The interconnections on the two levels are designed as fixed metallization patterns, and contact between the two patterns is made through holes etched in the second oxide layer.

Two-level Custom MSI.   Higher orders of complexity are usually accompanied by more specific custom requirements, and a second approach to MSI has been on a *custom-fabricated* basis.   The system uses several master slices.   On each master slice, a "cell" consisting of a mixture of selected gate circuits is repeated over the silicon slice.   The fabrication process is taken to the first level of metallization, which connects the elements into complete but separate gate circuits.   Then a second level of oxide is deposited all over the surface, and the slices are stored in inventory.   For any specific custom requirement, a second-level metallization pattern is designed as a fixed pattern to interconnect the number of gates and cells required to give the specified custom circuit.   The master slice is taken from inventory, holes are opened in the oxide to make contact to the gate contact pads, and the second-level interconnection pattern is deposited.

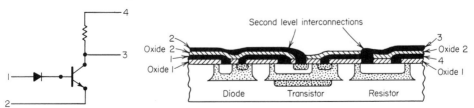

Fig. 8.4.   Two-level interconnection.

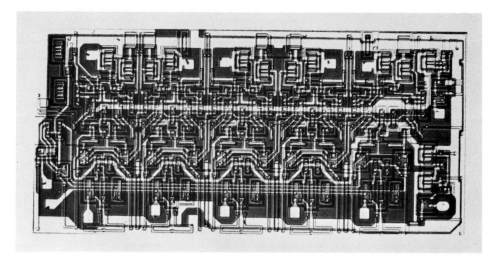

(a)

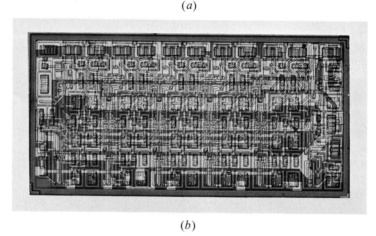

(b)

Fig. 8.5. Reduction in chip size by using two levels of interconnection: (a) single-level metallization, chip size 140 × 70 mils; (b) two-level metallization, chip size 110 × 55 mils.

To illustrate the method further, a particular master slice has 16 gates in the basic cell—a judicious mixture of input gates, internal gates, and output gates. This cell occupies an area of 60 mils square and is repeated over the whole slice. Now suppose a particular custom requirement calls for a 50-gate circuit. The second-level metallization pattern is designed to extend over four cells in a 2 × 2 matrix and makes contact to and interconnects 50 of the 64 gates. The slice is finally scribed up into chips 120 mils square, each chip containing the four cells with the second-level interconnection pattern. The general principle is illustrated in Fig. 8.6. Depending on the number of gates required in the final circuit, the number of cells to be used is chosen so that the slice can conveniently be scribed into chips. Obviously this method can be applied to any logic system, TTL, DTL, ECL, or MOS.

Hybrid Multichip MSI.   Another approach to MSI is the assembly of several

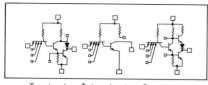

Input gate   Internal gate   Output gate

(a) First-level interconnection to complete gate circuits

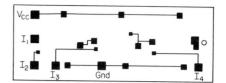

(c) Second-level metallization to interconnect gates into logic function

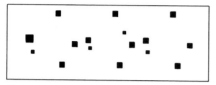

(b) Feed-through holes etched in second oxide layer

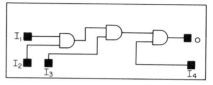

(d) Logic function

Fig. 8.6. Two-level custom MSI: general method.

chips onto a ceramic substrate which contains a thick-film pattern that interconnects the chips into the required functional circuit.   Each chip may contain up to the order of 10 to 20 gates, interconnected into convenient subsystems.   Figure 8.7 shows a four-chip assembly to give an overall 80-gate unit.   The wiring pattern on the ceramic can be formed by the silk-screen technique.   Two levels are used to allow crossovers on the ceramic by applying a glass slurry over the first level of connec-

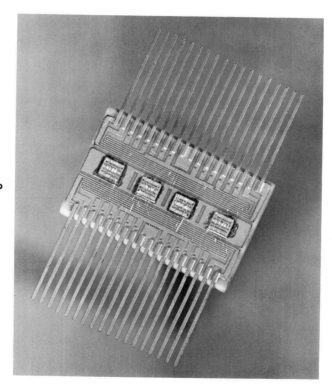

Fig. 8.7. Hybrid multichip MSI assembly.

tions and firing.   Then the second level of connections can be screened over the top.   The integrated-circuit chips are mounted onto prepared locations using several methods.   They can be mounted face up and connections made by ball bonding; they can be assembled face down so as to make all connections simultaneously directly to the chip (flip-chip assembly); or the chips can be fabricated with gold *beam leads,* and connections made by welding.   The assembled unit can be tested before final encapsulation, and any faulty chips replaced.

A distinct advantage of this approach is that standard integrated-circuit chips can be assembled to give medium levels of complexity.   Also, the arrangement can be very flexible in that combinations of different integrated-circuit chips can be used—for example, combinations of MOS and bipolar chips.

On the debit side are that special test procedures are necessary to identify faulty chips and that a high content of skilled labor is involved in both the initial assembly and, particularly, the rework to replace faulty chips.

MOS-MSI.   The master-slice approach described above for bipolar circuits can equally be applied to MOS-MSI circuits.   However, because of the relative simplicity of MOS circuit layout, with only the one basic MOS geometry, it is possible to design the layout of basic gates, shift-register bits, and flip-flops as standard cells and then use computer programs to arrange these cells into complete circuits, at the same time generating the artwork details for both the diffusion and metallization photomasks.   Using such computer-aided design methods, new designs both standard and custom can be prepared in a short time with a minimum of human errors. Because of the small area occupied by MOS structures, 100-gate MOS-MSI circuits can be fabricated on quite small chips—about 60 mils square.

LARGE-SCALE INTEGRATION

LSI Chip Technology.   Large-scale integration can first be considered as an extension of MSI techniques with what can be called LSI chip technology.   As the trend with MSI continues, and more circuits are included on a chip, the point will be reached where the arbitrary limit of 100 gates is exceeded, and then we will have LSI chips fabricated using MSI technology.   There will eventually be a limit to what can be included on a chip, but this limit will gradually move upward as improvements in optical resolution and photoresist processing allow element dimensions to approach the micron level.   Already with MOS circuits, complexities of several hundred gates are being achieved on relatively small chips.   With bipolar LSI chips, the use of two levels of metallization will generally be necessary.

LSI Hybrid Multichip Assemblies.   The second method of carrying out LSI is simply to use multichip hybrid assemblies of MSI chips having complexities in the range of 20 to 100 gates per chip.   By this means, complex assemblies of several hundred gates can be fabricated.   The general considerations are similar to those for MSI multichip assemblies mentioned above.

LSI Full-slice Technology.   LSI full-slice technology is aimed at realizing the full potential of integrated electronic components and is based on using the whole silicon slice as the packaged product.   This will allow very high complexity levels, up to several thousand gates in an IEC.   The procedure, being carried out so far with bipolar TTL technology, is to repeat several basic circuits, different types of gates and flip-flops, over the slice. The slices are processed in the routine manner up to the first level of metallization which interconnects the elements to form complete but separate circuits. So far, the only special feature is the supply of photomasks with intermixed circuit patterns. The circuits are now all tested on the slice by probe testing, and the location of good circuits is identified. Then, a computer is used to design an interconnection pattern to connect up good circuits to give the required system. This is called *discretionary wiring* and is carried out by using two additional levels of interconnection on top of the probe-tested slice, making three interconnection levels in all. Each interconnection level consists of a metallized pattern defined by photoresist techniques in the usual manner. A layer of insulation such as silicon oxide is formed between each two levels of metallization. Each interconnection in the discretionary wiring is arranged as two components at right angles. All the horizontal components are formed with the second level of metallization, and the vertical components with the third. The two components are linked together through holes in the intervening oxide.

Faulty gate circuits are randomly distributed from slice to slice, and so every slice will have a different and unique discretionary-wiring pattern. In practice, each slice is probe tested automatically, and the data on the location of good circuits are fed to a computer together with the specification of the overall system required on the slice. The computer is programmed to develop the routing of the discretionary interconnections between the good circuits, the output data being in a form that will allow the fabrication of the necessary photomasks for opening holes in the oxide layers and for forming the required metallic interconnections.

A series of master slices is again used to allow the fabrication of different functions. Each master slice contains a different distribution of basic circuits. A typical slice is shown in Fig. 8.8, and an enlarged view of part of the slice in Fig. 8.9. The slice contains 112 single-input gates, 180 three-input gates, 112 five-input gates, 22 seven-input gates, and 95 flip-flop circuits. At probe test the circuit yield is typically between 70 and 80 percent, making a total of about 300 good gates and 60 good flip-flops available for discretionary interconnection. Figure 8.10 shows the slice after the second-level metallization, and Fig. 8.11 after the final third level. The completed slice is mounted to a special 160-lead header as in Fig. 8.12 and encapsulated to give a unit about 2 in. square, as shown in Fig. 8.13.

The potential of this approach is such that the main central unit of an average, general-purpose computer could be built using 30 to 40 LSI full-slice units. Inevitably, the cost of setting up the fabrication facilities for discretionary wiring is quite high. Also, at the present stage of development, the cost of determining the routing for random logic systems is also high. The main technical problem is in maintaining high yield of the multilevel interconnection system.

It would appear that active memory stores and serial shift registers are particularly suited to fabrication by full-slice LSI because of the relative simplicity of the interconnections between circuit bits. With the memory store, a simple matrix of horizontal and vertical connections is required, and with shift registers, the connections are serial, from one bit to the next. It is possible to introduce fairly simple redundancy to allow for inoperative bits, and then the interconnections can be carried

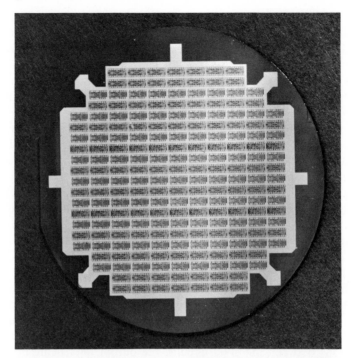

Fig. 8.8.   LSI master slice with intermixed gates and flip-flop circuits (slice diameter 1.25 in.).

Fig. 8.9.   Enlarged view of part of slice in Fig. 8.8.

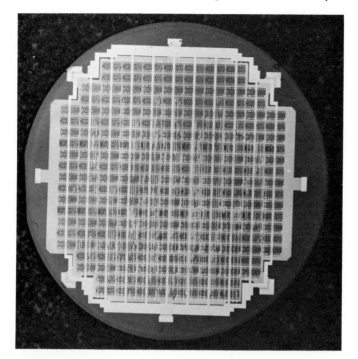

Fig. 8.10. LSI slice after second-level interconnection.

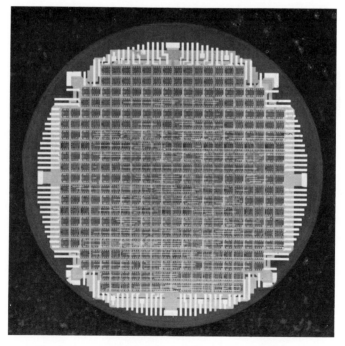

Fig. 8.11. LSI slice after third-level interconnection.

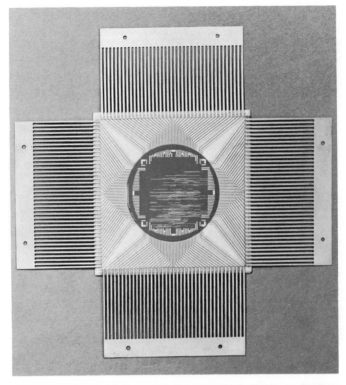

Fig. 8.12. Mounted LSI slice. The lead frame straps are sheared off after assembly to separate the leads.

Fig. 8.13. Encapsulated LSI slice.

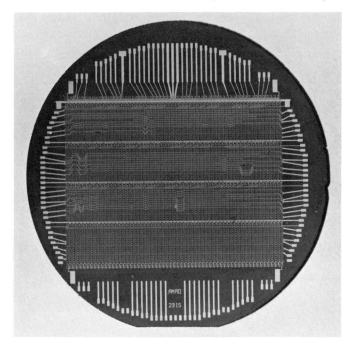

Fig. 8.14.   LSI memory slice (slice diameter 1.25 in.).

out with only two levels of metallization.   Figure 8.14 shows a completed memory slice containing a total of 1,024 memory bits (32 words each, up to 32 bits).

MOS Full-slice LSI.   Because of the high complexity that is possible with MOS circuits, MOS-LSI technology has mainly been at the chip level.   With MOS random logic, although several hundred logic gates can be formed on a small chip, the problem of routing interconnections becomes very difficult.   However, as with the bipolar case described above, there are interesting possibilities for MOS full-slice memory arrays.   One MOS memory bit can be formed in about 50 square mils of area, and so a matrix array of up to 10,000 such bits can be formed on a single slice of silicon 1.25 in. in diameter.   Thus there is the possibility of fabricating very high capacity memory arrays using MOS-LSI full-slice techniques.   The use of redundancy and discretionary wiring will apply as for bipolar LSI described above.

CONCLUSIONS

In this lesson, we have looked a little into the future.   Standard-chip MSI with complexity up to about 35 gates is now routine manufacture, and custom MSI is being built up into a quick turnaround facility.   With both of these, the complexity will gradually increase toward the 100-gate level.   The principles and basic approach of full-slice LSI have been established, and activity is now directed to engineering the technology into a manufacturing process.   In both the MSI and LSI areas, hybrid multichip technology shows its possibility as an alternative approach.

Thus, as integrated circuits continue to take over from discrete devices, integrated electronic components will gradually assume greater importance, with MSI-IECs building up steadily and specialized applications of full-slice LSI appearing in the near future. There can be no doubt that MSI and LSI in their various forms will result in IECs that will play a very important part in future electronics.

GLOSSARY

complexity   The number of equivalent gates in a circuit.

discrete device   An individually packaged semiconductor-device element.

discretionary wiring   A method of interconnecting into a complete functional system only the good integrated circuits on a silicon slice, which may contain both good and inoperative circuits randomly located.

full-slice technology   A system in which a complete electronic system is formed on a single full slice of silicon, which is then packaged as a single unit. The process may include discretionary wiring to interconnect the basic circuits.

integrated circuit (IC)   An electronic circuit which has been fabricated as an inseparable assembly of semiconductor elements in a single chip of silicon with a circuit complexity less than 10 equivalent gates.

integrated electronic component (IEC)   A complete, functional electronic subsystem formed by interconnecting several integrated circuits on a single chip of silicon and assembling it into a single package, to give a unit with a circuit complexity greater than 10 equivalent gates.

large-scale integration (LSI)   The group of technologies used in the fabrication of integrated electronic components with circuit complexities above 100 equivalent gates.

master slice   A partly processed silicon slice, containing a selection of complete but separate gate circuits. These gate circuits can then be interconnected in various ways by subsequent processing, to allow the fabrication of custom functional subsystems.

medium-scale integration (MSI)   The group of technologies used in the fabrication of integrated electronic components with circuit complexities in the range of 10 to 100 equivalent gates.

REVIEW

For each of the numbered statements below, select the one of the items lettered *a, b, c,* or *d* that correctly completes the statement.

8.1. An integrated electronic component is
   a. One circuit element of an integrated circuit.
   b. A discrete device assembled into a hybrid circuit.
   c. An electronic component included in an electronic circuit assembled on a printed circuit board.
   d. A complete electronic function, consisting of several integrated circuits formed and interconnected on a single chip of silicon with a circuit complexity above 10 gates.

8.2. If a chip with more than two or three gates is assembled into an integrated-circuit package
   a. The total cost is the same.
   b. The packaging cost is shared between more gates, resulting in a lower cost per gate.
   c. The system cost will be higher.
   d. The system reliability will be worse.

8.3. If an integrated-circuit chip is increased in area to allow a more complex circuit,
   a. The number of good chips per slice will remain the same.
   b. The increase in chip area will be proportional to the increase in complexity.
   c. The yield of good chips on the slice will decrease.
   d. The yield of good chips will increase.

8.4. As the complexity of a packaged integrated circuit is increased
   a. The chip cost per gate decreases.
   b. The packaging cost per gate remains the same.
   c. The overall cost always increases proportionally to the complexity.
   d. The overall cost per gate passes through a minimum value at some level of complexity.

8.5. When using integrated circuits in an electronic system
   a. It is most economical to use types with the optimum complexity in order to obtain the lower possible cost per gate.
   b. It is always best to use simple-gate types mounted on a printed circuit board.
   c. The total system cost will be the same whatever type is used.
   d. The cost of printed circuit boards is negligible.

8.6. The term medium-scale integration refers
   a. To integrated circuits with complexity below 10 gates.
   b. To integrated circuits manufactured on a medium-volume production line.
   c. To the technologies used in the fabrication of integrated electronic components with complexities in the range of 10 to 100 gates.
   d. Only to those integrated circuits which have a single level of metallization.

8.7. Medium-scale integration
   a. Often uses two levels of metallization.
   b. Can only be used for custom types.
   c. Only applies to single-chip circuits.
   d. Can only be applied to one logic system.

8.8. The two-level system of interconnection
   a. Requires an increased chip area.
   b. Can be used to advantage on custom MSI units using master slices.
   c. Must not allow interconnections on the two levels to cross over each other.
   d. Can never be used for integrated electronic components using LSI.

8.9. Large-scale integration
   a. Refers only to IECs fabricated on full slices of silicon.
   b. Cannot be applied to MOS types.
   c. Refers to technologies used to fabricate any IEC with circuit complexity above 100 gates.
   d. Does not require costly process facilities.

8.10 Large memory arrays
   a. Cannot be formed using MOS techniques.
   b. Can be made using full-slice LSI techniques including redundancy and two-level interconnection.
   c. Use a large number of circuit bits, all connected in series.
   d. Are not suitable for fabrication by full-slice LSI techniques.

LESSON 9

The Application of Integrated Circuits

INTRODUCTION

Looking back over the past few lessons, it will be seen that integrated circuits have been developed and are generally available in the following categories:

1. *Digital circuits*—suitable for use in general-purpose digital computers, digital measuring instruments, digital control systems, and memory systems.
2. *High-speed digital types* (ECL)—for use in high-speed data processing.
3. *Operational amplifiers*—for linear and analog functions in the range from dc up to several tens of megahertz.
4. *Differential amplifiers*—for comparator applications, particularly in industrial control systems.
5. *Sense amplifiers*—for interfacing between linear systems and digital systems.
6. *Audio, video, and high-frequency linear amplifiers*—applicable to communication and consumer entertainment equipment.
7. *Microwave integrated circuits*—of the hybrid type for application in radar and microwave communication links.

In addition, special-purpose circuits are being developed incorporating optoelectronic combinations, thermal coupling, and transducer-amplifier combinations.

From this range of circuits, it is clear that integrated circuits can now be applied to a wide variety of electronic systems. It has been forecast that by 1978, between 85 and 95 percent of all electronic functions will be carried out by integrated circuits or integrated electronic components. The great motivation to use integrated circuits is the attractive combination of lower overall cost and improved reliability.

The main functions that have not yet been "integrated" are in the power area. But there can be little doubt that power-device structures will be combined with associated control devices, initially as hybrid multichip assemblies in one package and eventually as complete monolithic units.

The application of integrated circuits is not just a question of replacing existing electronic circuits with the same circuits fabricated in integrated form, as tended to be the case when transistors began to replace vacuum tubes. The new possibilities brought along by integrated circuits are resulting in design engineers' going back to basic thinking and evolving new concepts of electronic manipulation. We have entered a new era—the integrated circuit era of electronics—and the next few years will see a profound change in most electronic equipment.

Since, as mentioned above, integrated circuits can and soon will be used in the majority of electronic equipment, a full discussion on the application of integrated circuits would be a voluminous presentation covering virtually all electronic equipment. Obviously, this is not the intent of this course. The aim in this lesson is to give a general appreciation of *where* integrated circuits are already being used and to indicate other areas where they will doubtless be applied with advantage in the near future.

PACKAGING OF INTEGRATED CIRCUITS

Before discussing the application of integrated circuits in electronic systems, it will be well to briefly review how they are assembled into compact subsystems for use in the overall equipment. The alternative forms of integrated-circuit packages were mentioned in earlier lessons. They are the multilead TO-5 circular package, the flat pack, and the dual-in-line (both plastic and ceramic). They are shown again in Fig. 9.1. The multilead TO-5 package was widely used with early types of integrated circuit, mainly because it was readily available, and equipment for using it was established for transistor fabrication. However, it has disadvantages from the equipment-assembly viewpoint; the lead spacing is very small, and this introduces problems in printed-circuit-board assembly. The flat pack is advantageous where small size is important, but requires relatively costly equipment for assembly onto boards. The dual-in-line package was specifically designed to be compatible with standard printed-circuit-board assembly methods and, as a result, is proving to be

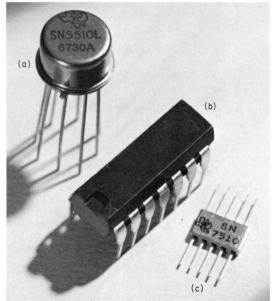

(a)

(b)

(c)

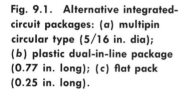

Fig. 9.1. Alternative integrated-circuit packages: (a) multipin circular type (5/16 in. dia); (b) plastic dual-in-line package (0.77 in. long); (c) flat pack (0.25 in. long).

very popular. It would appear that it will continue to become even more popular as time goes on.

Most integrated circuits are available in all package versions. Because of the features mentioned above, when selecting the package type, the choice should include consideration not only of the cost of the packaged unit itself, but also of the cost of assembling it onto a printed circuit board.

PRINTED-CIRCUIT-BOARD ASSEMBLY

The standard practice in equipment assembly of integrated circuits is to use double-sided printed circuit boards with through plated holes to interconnect the circuit wiring on the two sides. This system allows interconnection conductor crossovers, with conductors on either side of the board. With the flat-pack and dual-in-line packages, the "bridging" of conductors can be achieved by the disposition of the package itself, but this is not possible with the TO-5 package because of the close spacing and circular arrangement of the leads.

The preferred method of mounting the flat pack is to weld the leads straight down to conductors on the top surface of a printed circuit board. A precision parallel-gap welder has been developed for this purpose, with two electrodes having tips 20×15

Fig. 9.2. An assembly of flat-pack integrated circuits on a single-sided printed circuit board.

Fig. 9.3. An assembly of dual-in-line integrated circuits on a double-sided printed circuit board.

mils separated by only 20 mils. This system requires special metal such as nickel or plated kovar on the printed circuit board to allow good welding. The merits of this single-sided assembly are that no manipulation of the flat-pack leads is necessary, and the assembly processes are all carried out on one side only. The flat pack with welded assembly is particularly used for high-reliability military applications. Figure 9.2 shows a completed board with flat-pack units welded in position.

By bending the leads down through 90°, the flat pack can be inserted into holes in a double-sided board and flow soldered, but this method results in very close spacing between the solder pads.

The dual-in-line package has a lead spacing of 0.1 in. between centers, and the leads project downward so that the unit can readily be inserted into a standard printed circuit board and flow soldered on the back surface. Figure 9.3 shows a typical board assembly using the dual-in-line package. A particular advantage of the dual-in-line arrangement is that conductors can run along between the two rows of pins, allowing the possibility of simpler routing of connections with consequent saving in board area.

Printed circuit "boards" (or "cards") are generally between 6 and 10 in. square and are assembled in rack form as illustrated in Fig. 9.4.

EQUIPMENT DESIGN USING INTEGRATED CIRCUITS

A number of important decisions must be made when considering how to use integrated circuits in equipment.

With digital equipment, a family of logic integrated circuits must be chosen and

Fig. 9.4. A rack assembly of integrated-circuit boards.

used throughout the system.   In general, most digital systems can now be designed completely with integrated circuits, and so the printed-circuit-board assemblies do not include any discrete components.   Because new types of digital integrated circuits tend to be more complex, with more gates in one package at a lower cost per gate, the content of a printed circuit board should be designed so that it will be possible to take advantage at a later date by substituting a simpler card containing a smaller number of more complex circuits or even by substituting a single LSI unit.

With linear integrated-circuit applications, it is usually necessary to include some discrete components; for example, operational amplifiers will need external negative feedback and compensation components, and tuned i-f amplifier stages will require external tuned circuits.   In such cases, the dual-in-line package is very advantageous, as the discrete components and the integrated circuits can readily be assembled onto a standard printed circuit board in the same operation.

The relative price of various integrated circuits is, of course, a major factor in deciding which type to use.   Because of the dependence of price on the volume of manufacture and the tendency for prices to fall with time, it is important to base the decision on the price that will apply to production quantities at the time they will be required for production runs, which may be lower than the current price.

FIELDS OF APPLICATION

Integrated circuits are already being used for about 25 percent of all electronic circuits. By 1978 this figure will have increased to about 90 percent. It is of interest to see where integrated circuits are being applied now and the possibilities for their application in the immediate future.

The major fields of application can be stated as follows:

Computers.
Desk calculators.
Industrial control.
Aircraft electronic equipment.
Space vehicles and missiles.
Automobile electronic equipment.

Electronic instruments.
Communication and telephone.
Consumer entertainment.
Consumer appliances.
Medical electronics.

The application of integrated circuits in each of these categories will be briefly discussed to indicate the functions for which they are being, and soon may be used.

Computers.   Solid-state devices have successively had a major influence on the trend of electronic computers. This is illustrated in Table 9.1, which gives some statistics of four generations of computers, starting with a vacuum-tube type and progressing through discrete-transistor and integrated-circuit types to a comprehensive equipment being developed with integrated electronic components. The first three computers were designed and built for seismic applications, and the fourth will be an advanced scientific computer using MSI-LSI techniques. Figure 9.5 shows an artist's conception of what the latter will look like. It will be seen that integrated circuits allowed a significant improvement in the number of gates per unit equipment volume over the discrete-transistor type, and the use of high-complexity integrated electronic components will increase the improvement much more.

Discrete devices are no longer being used for any digital computer development. While there may be some divergence of opinion between computer manufacturers on which type of logic circuit and package to use, there is universal agreement on the use of integrated circuits. Digital computer circuits are designed using standard circuits of the general type discussed in Lessons 5 and 7. Analog computers are

Table 9.1.   Progress in Computer Development

Computer type, Texas Instruments	Year	Active element	Number of gates	Equipment volume, cubic feet	No. of gates per cubic foot
DARC	1957	Vacuum tube	300	100	3
827	1960	Transistor	1,500	75	20
870	1966	Integrated circuit	4,500	50	90
Advanced scientific computer	1972	Integrated electronic component	220,000	130	1,700

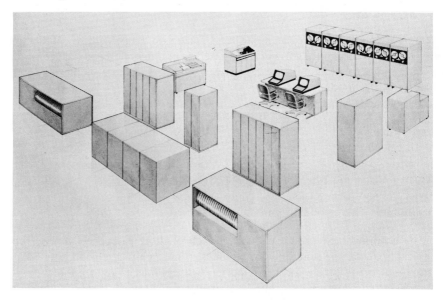

Fig. 9.5. An artist's impression of an advanced scientific computer system.

designed around the operational amplifier and other standard circuits and, as mentioned above, must include some discrete components in the feedback networks.

The development of integrated-circuit active-memory arrays, discussed in Lesson 8, will have a significant influence on the future design of computers. In the past, ferrite-core memories have been used as separate central units. This is not the most efficient arrangement with large computers, and we may see smaller, very fast, integrated active memories distributed at strategic points in the logic systems.

Desk Calculators. Up to the present time, desk calculators have been based on mechanical or electromechanical operation. The cost of electronic systems using discrete-transistor assemblies was too high to allow economic electronic operation, but integrated circuits have changed this and are now being used in new calculator designs such as the type illustrated in Fig. 9.6. Thus, we have an interesting case where integrated circuits have made possible something that was not economic using discrete components. Because the operating speed requirement for this type of calculator is only moderate, high-complexity MOS integrated circuits will probably be used extensively.

Industrial Control. Integrated circuits are being applied to a wide variety of equipment in the field of industrial control, and this promises to be one of the most important fields of application. The applications are varied, but standard integrated circuits are being used successfully, resulting in attractive low costs. General applications such as voltage control, temperature control, speed control, power conversion, and power-overload protection are using integrated circuits in conjunction with silicon controlled rectifiers and Triacs. Custom applications for specific requirements are now being developed in increasing numbers, and mention of a few equipments already using integrated circuits will indicate the spread of activity.

1. *Machine tool control*—for numerical, position, and contour control (Fig. 9.7).
2. *Papermaking control*—for measuring the degree of shrinkage or stretch of paper between rollers and applying feedback correction.
3. *Multipoint monitoring and recording*—to monitor up to 1,000 points in power plants and industrial systems.
4. *Road-bed volume computing*—to calculate the volume of cement, crushed rock, and asphalt required for new roads.
5. *Food sorting*—to inspect and sort food, such as peas and corn, at a rate up to 25,000 units per minute.
6. *Water metering*—for automatic recording and billing of water supplies.
7. *Bridge crane control*—to give complete control of hoist, trolley, and bridge movement.

These are but a few of the many cases where integrated circuits are being used in industrial control. Most of the applications started as discrete-component assemblies, and now the use of integrated circuits is resulting in better performance and higher reliability at lower cost.

Because of the wide range of applications in the industrial-control field, this topic will be discussed in more detail in the last lesson of the course.

Aircraft Electronic Equipment.   Practically all electronic functions in aircraft can

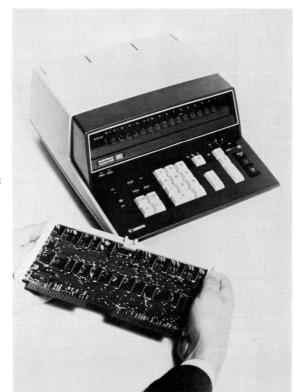

Fig. 9.6. A desk-top electronic calculator with the main integrated-circuit assembly board.

Fig. 9.7. A contour-milling machine controlled by electronic equipment (on the left) incorporating integrated circuits.

be carried out using integrated circuits. It is estimated that at present about 25 percent of the electronic equipment on new aircraft uses integrated circuits, and this should increase to 50 percent within the next two to three years. Although the initial incentive to use integrated circuits in aircraft equipment was the small size and weight, considerable emphasis is now on reliability. Electronic equipment on aircraft is continually tending to become more complex as more systems are included, and so low cost will always be an important consideration. To name only some of the equipments in which integrated circuits are—or soon will be—used, we have radio communication and navigation systems, radio altimeters, autopilot and beam approach equipment, radar sets, and power-control systems. On a modern military aircraft, several thousand integrated circuits are already used, and the number is increasing continually.

Space Vehicles and Missiles. It almost goes without saying that integrated circuits are being used wherever possible in the electronic equipment for space vehicles and missiles. As the use of orbiting satellites for communication and the like increases, so will the use of integrated circuits, both in the satellites and in the ground-control equipment.

Automobile Electronic Equipment. Until recently, the only use of electronics in the automobile was the radio receiver. The low cost and high reliability of mass-produced integrated circuits are now prompting developments in a number of possible applications. Several of these are concerned with engine control. One of the first functions that is being converted to use an integrated circuit is the voltage regulator. The integrated circuit will be mounted inside the alternator together with the silicon rectifiers to give a complete self-contained unit. Electronic fuel control and ignition systems using discrete-component assemblies have already been proved to some extent, and integrated-circuit versions are being developed. Engine revolution counters (tachometers) and road speedometers are both suitable for integrated-circuit operation, and the latter could be combined, perhaps in one integrated circuit, with an electronic gasoline-metering system to give a gasoline-consumption meter calibrated directly in miles per gallon. Existing automatic transmission systems can be improved with simple electronic computer control—a natural application for integrated circuits.

A second group of integrated circuits will be associated with automobile safety and will include such functions as antiskid control, speed regulation, vehicle proximity indication, improved headlight dimming, safety interlocks, and fault-indicator systems. Taken individually, these various applications are not yet attractive economically, but the possibility of combining several to use one or two integrated circuits may present an earlier solution.

The third area for integrated circuits is in comfort, entertainment, and communication systems. The existing solid-state radio and tape player will convert to integrated circuits as economics allow, and they may be extended to include some form of automatic road-hazard communication from outside. There is a need for better temperature control; integrated circuits may help here.

Integrated circuits, like automobiles, are essentially suited to mass production. Although the use of integrated circuits in automobiles is only just starting, the wide possibilities mentioned above will doubtless give the incentive for a fast buildup over the next five to ten years if the cost can be reduced to a sufficiently low level.

Electronic Instruments.   Integrated circuits are already being used in such electronic instruments as digital voltmeters and decade counters. These applications can use techniques developed for computer applications with little change. The use of standard digital integrated circuits is allowing higher precision and stability with more comprehensive operation at a lower overall cost. On the linear side, operational amplifiers with feedback to give stable and accurate "gain-of-10" and "gain-of-100" amplifiers are being used as the basis of multirange instruments, and differential amplifiers are being applied in comparator equipments.

For measuring instruments, the high reliability of integrated circuits is perhaps the most important characteristic, with the low cost always a welcome feature. The low operating voltage and low power consumption allow operation from an internal battery, if desired, to give a "floating" condition with isolation from the supply mains. The low power consumption is also advantageous in that it results in only a low temperature rise inside the equipment and helps to give better maintenance of calibration.

Communications and Telephone.   Rf communication equipment covers the 2 to 76-Mhz frequency band. Prototype receivers have been designed around a family of four integrated circuits: an r-f amplifier, an i-f amplifier, an audio amplifier, and an AGC circuit. These four integrated circuit units are linked with external tuned circuits, a mixer and local oscillator, and a second detector unit. A block diagram of the general arrangement is shown in Fig. 9.8. Up to now, communication equipment has generally consisted of linear systems. With the availability of low-cost digital integrated circuits, digital methods of signal processing are being investigated and may result in more economical arrangements.

For telephone communication, integrated-circuit audio amplifiers have been designed for microphone, headphone, and line amplifiers. These will be used to increase the audio level where necessary and to improve the signal-to-noise ratio. Telephone-exchange systems are being designed to use pulse-code modulation and time-division multiplex in order to use multichannel telephony with a minimum of lines. For this purpose, MOS integrated circuits are finding use for operations such as cross-point matrix arrays to switch the digital signals through the exchange to the required destination.

Consumer Entertainment. The application of integrated circuits to radio, television, and audio systems is inevitable, it is just a question of taking turn in the development program to design the required circuits and produce them at a sufficiently low cost. Up to the present time, only one or two isolated instances of application have been reported, such as the sound i-f amplifier in a television receiver and an experimental radio.

Low-cost record players will probably use a single integrated-circuit audio amplifier between the pickup cartridge and the loudspeaker, with an output of a few watts. Audio systems will incorporate integrated circuits as microphone and phonograph preamplifiers and in low-level driver amplifiers. Radio receivers will initially use integrated circuits in the i-f and audio amplifiers.

With regard to television receivers it has been said, "What the transistor did for the radio receiver, the integrated circuit will do for the television receiver." Because of its complex nature, the television receiver will be a hybrid system using as many integrated circuits as possible, complemented by discrete-transistor circuits for such functions as deflection output. Figure 9.9 shows a block diagram how a color television will probably be designed to use integrated circuits. All functions to the left of the dashed line will be low-level circuits carried out by integrated circuits. The color and deflection output circuits will use power transistors.

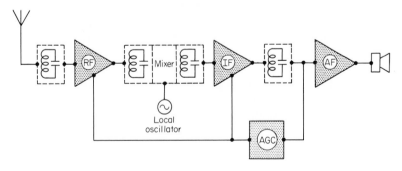

Fig. 9.8. Block diagram of integrated-circuit radio receiver.

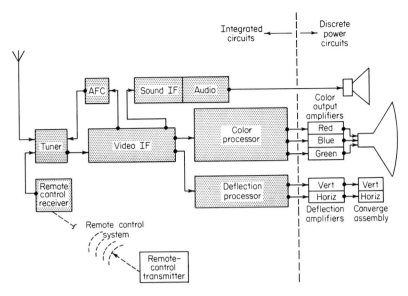

Fig. 9.9. Block diagram showing how integrated circuits will be used in television receivers.

Audio tape and video tape equipments will also go over to use integrated circuits in the near future. In particular, the use of integrated circuits in video tape units may have a major impact and bring the economics down to a level suitable for the domestic market.

Consumer Appliances.   There is a continual trend for domestic appliances to become more versatile. Washing machines, clothes driers, dishwashers, and cookers all have multicycle operation which at present is carried out by a mechanical selector switch driven by a small synchronous motor. The feasibility of carrying out such operations electronically has been demonstrated, using discrete-transistor circuits coupled to silicon controlled rectifiers or Triacs. Up to now, the cost of the electronic approach has been too high, and the electromechanical system will continue until the cost of an integrated-circuit–Triac combination falls to a sufficiently low value.

Medical Electronics.   Although microelectronics would appear to be ideally suitable for many medical uses, developments of such applications have been relatively slow, presumably due to lack of funding.

The use of the computer for record analysis and diagnostic routines is now building up. Integrated circuits are beginning to replace discrete devices in implanted units such as heart pacemakers and in nerve stimulators. Considerable work has been carried out in the space program on astronaut monitoring, and integrated-circuit systems based on this work will doubtless soon be utilized in hospitals for patient monitoring systems.

And one should not forget the hearing aid, the very first commercial application of transistors. Modern hearing aids of the in-the-ear or behind-the-ear types use a single integrated circuit audio amplifier.

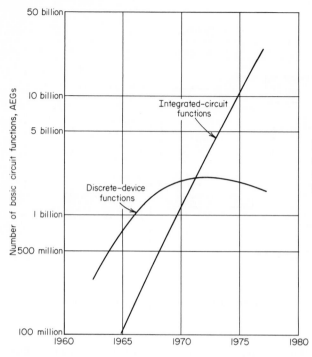

Fig. 9.10. Trends of the numbers of basic circuit functions in the United States.

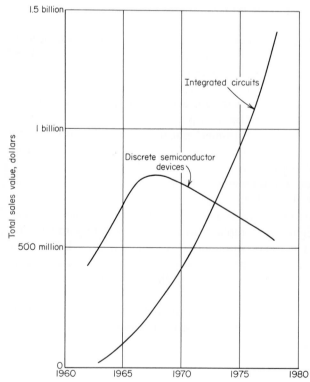

Fig. 9.11. Trends of discrete-semiconductor-device and integrated-circuit sales value in the United States.

CONCLUSION

The foregoing paragraphs will have indicated how the application of integrated circuits is building up across the whole spectrum of electronic equipment. To present an overall picture, Fig. 9.10 shows the trends of discrete-semiconductor-device (transistors and diodes) and integrated-circuit (including integrated electronic components) usage in electronic equipment in the United States. The unit of the vertical scale is the smallest functional group of circuit elements into which electronic equipment can be subdivided, such as a single amplifier stage. This normally contains one active element, such as a transistor, and is called an *active element group* (AEG). The curve shows the rapid buildup of integrated-circuit application and how it is expected to continue during the next ten years. The cost of integrated circuits will steadily decrease as techniques improve. The dollar values of discrete devices and integrated circuits corresponding to the quantities of AEGs shown in Fig. 9.10 are plotted in Fig. 9.11.

GLOSSARY

active element group The smallest functional group of circuit elements into which an electronic equipment can be subdivided, e.g., the capacitors, resistors, etc., plus an associated transistor in a one-transistor amplifier stage.

active-element memory A memory circuit using active devices such as transistors to establish the memory states.

parallel-gap welder A system of welding in which the two electrodes are positioned side by side and are applied together to the surface of one metal which is to be welded to a second metal underneath. The welding current flows down from one electrode, spreads out through both metals, and then flows back up to the second electrode.

printed circuit An electric circuit in which the wiring is formed by "printing" the required pattern onto a metal coating on the surface of an insulating sheet and etching away the unwanted metal.

REVIEW

For each of the numbered statements below, select the item lettered *a, b, c,* or *d* that correctly completes the statement.

9.1. Digital integrated circuits are designed for application in
   a. Analog industrial-control equipment.
   b. Voltage comparator systems.
   c. Logic systems for digital computers.
   d. Audio amplifiers.

9.2. The application of integrated circuits
   a. Will only take place in computers.
   b. Will not involve new concepts.
   c. Consists of replacing a discrete-component circuit with the same circuit fabricated in integrated form.
   d. Will cause a profound change in most electronic equipment.

9.3. One advantage of the dual-in-line package is that it
   a. Is the smallest package.
   b. Was readily available when integrated circuits were first made.

      c. Has very close lead spacing.

      d. Is compatible with printed-circuit-board assembly methods.

9.4. The dual-in-line package is usually assembled onto printed circuit cards by

      a. Using flow-soldering techniques.

      b. Welding.

      c. Soldering each lead individually

      d. Using parallel-gap soldering.

9.5. When designing equipment using integrated circuits

      a. It is usual to use a combination of different logic types.

      b. One should never consider incorporating any discrete components.

      c. It is sensible to make provision where possible to allow future substitution of more complex types having a lower cost per gate.

      d. It must be remembered that the prices of integrated circuits tend to rise with time.

9.6. Digital computers

      a. Have not been influenced significantly by the development of integrated circuits.

      b. Are generally designed so that a particular computer uses only one type of logic circuit throughout.

      c. Are still being developed using discrete components.

      d. Have always used transistor memory circuits.

9.7. The application of integrated circuits to industrial control

      a. Will not be economic for custom applications.

      b. Has not yet taken place.

      c. Promises higher reliability at a higher cost.

      d. Has already been carried out successfully in a variety of equipments.

9.8. The application of integrated circuits in measuring instruments

      a. Will only involve linear types.

      b. Will take advantage of their low operating voltage and low power consumption.

      c. Is not feasible from the stability viewpoint.

      d. Will require the development of special types of circuit.

9.9. In the consumer field, the application of integrated circuits

      a. Is not yet being carried out on a production basis.

      b. Will be restricted to entertainment systems.

      c. Holds no promise.

      d. Is already widely carried out.

9.10. MOS integrated circuits are being developed for

      a. High-speed data processing.

      b. Microwave circuits.

      c. Matrix switching arrays in telephone exchanges.

      d. The very-high-frequency circuits of r-f communications receivers.

The Use of Integrated Circuits
in Electronic Control

GENERAL CONSIDERATIONS

In any industrial equipment using electric power, there is the need to control the power supplied to the load. First, the power must be switched on and off in order to use the equipment. With certain types of load this is not as easy as it might seem. Second, in many equipments, variation of power is required to allow operational use; for example, to give speed variation of motors, temperature control of furnaces, and brightness variation of lamps. Such functions often require that the electrical power supplied to the load can be varied smoothly from 0 to full power. Third, much equipment is required to operate continuously under controlled constant conditions—constant speed, constant temperature, constant brightness. In this case, the input power needs to be controlled automatically by feedback signals such that the constant conditions are maintained at all times.

A block diagram of a generalized power control system is shown in Fig. 10.1. The power supply is fed from the supply mains through the power control system to the load, and various feedback arrangements are included to allow variable or constant output conditions as required. Solid-state devices are now available to carry out all the control functions. For the main power control, a class of device called *thyristors* have been developed. There are two main types, the *silicon controlled rectifier* (SCR) which controls the flow of current in one direction and so allows dc control and the *Triac* which controls the flow of current in both directions to allow ac control. Both of these devices are triggered by small current pulses fed from a driver stage, control of the output power being arranged by varying the phase of the pulses relative to the supply. The trigger pulses are, in part, derived from signals generated by a transducer associated either with the load itself or with some external consequence of the load. These signals are amplified and fed back to the power control system so as to give the required control. It is in the manipulation of this feedback signal—amplifying it, comparing it with a reference, mixing it with sequence control programs, and generating driving impulses—that integrated circuits are starting to play a very important role in industrial electronic control.

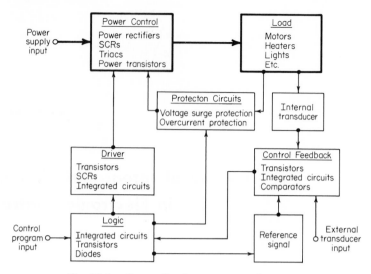

Fig. 10.1.   Generalized power control system.

To illustrate further the basic control system, consider the case when the load is a furnace which is required to operate at a predetermined constant temperature. The transducer in this case is a thermocouple located in the furnace hot zone. The output signal from the thermocouple is connected to a control circuit where it is compared with a reference signal, and the difference is amplified and fed to a logic circuit. Here, any other required programs are introduced (such as time cycles) and the resulting output is fed to the driver circuit where the final trigger pulses are generated. If the temperature is required to vary according to some predetermined law, the logic circuit can be arranged to vary the reference voltage, resulting in a change in the feedback difference signal. Manual control of the temperature can be carried out either by direct variation of the driver pulses or by variation of the reference signal.

Another example might be a street lighting system, to be switched on when daylight has fallen to a given level as indicated by an external transducer (a photocell) or at a given time, whichever happens first, as determined by a simple logic circuit.

In order to see how integrated circuits are being used in such industrial electronic control systems, it is first necessary to have some knowledge of the power control devices at one end of the system and of transducers at the other end.

THE SILICON CONTROLLED RECTIFIER AND DC POWER CONTROL

The silicon controlled rectifier (SCR) is a rectifying device, which normally blocks flow of current in both directions, but can be triggered into a condition that allows current flow in the forward direction while still blocking in the reverse direction. It is a semiconductor device with a p-n-p-n structure and has characteristics as illustrated in Fig. 10.2. The reverse characteristic is very high resistance, similar to that of a conventional silicon power rectifier. The forward resistance is initially high and blocks forward current, but if a small trigger pulse is applied to a third

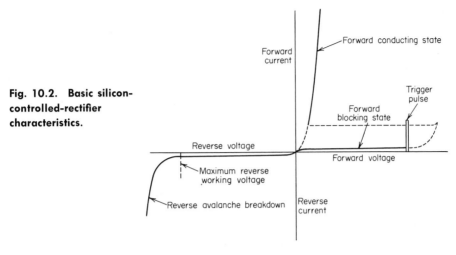

Fig. 10.2. Basic silicon-controlled-rectifier characteristics.

electrode, called the *gate,* the device changes to a low resistance and allows forward current to flow. Once the device is triggered, it remains in the conducting state until the load current is reduced almost to 0, when it automatically returns to the high-resistance blocking state.

Structurally, the SCR consists of a silicon wafer with a p-n-p-n arrangement. It can be considered as a combination of a p-n-p transistor and an n-p-n transistor with a common-collector junction between the center n-p junction.

The basic method of using the SCR for dc control from an ac main supply is illustrated in Fig. 10.3. Consider a sinusoidal voltage applied to the SCR. During the reverse half cycles, the SCR blocks, and no current flows. As the voltage increases in the forward direction, no current flows until a trigger pulse occurs.

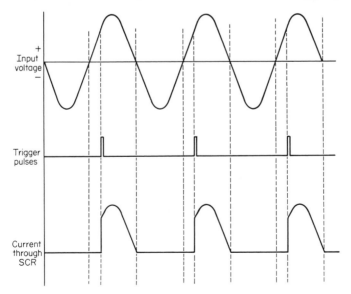

Fig. 10.3. Basic operation of the SCR.

Then the device triggers to a low-resistance state, forward current flows and follows the sinusoidal voltage for the remainder of the forward half cycle. At the end of the half cycle, the device returns to the blocking state and remains so until another trigger pulse is applied during the next forward half cycle. By varying the timing of the trigger pulses, relative to the supply waveforms, the proportion of the forward half cycle during which conduction takes place can be varied, and so the mean dc rectified current can be controlled. This is called phase control and is carried out by including phase shifting networks in the driver circuit.

THE TRIAC AND AC POWER CONTROL

The Triac is a further development of the SCR which can be triggered into conduction in both forward and reverse directions. It is functionally equivalent to two SCRs connected in reverse parallel with a common gate, and it consists of a silicon wafer with an n-p-n-p-n structure. The single gate input is arranged so that the device can be triggered into conduction by either positive or negative pulses with either polarity of the main-terminal voltage.

The method of using the Triac for ac power control is illustrated in Fig. 10.4. As the supply voltage increases in the forward direction, no current flows until a trigger pulse occurs and triggers the device into forward conduction. The device then remains conducting until the end of the positive half cycle, when it returns to the high-resistance blocking state and remains so until the next trigger pulse occurs during the negative half cycle. Reverse current then flows and continues to the

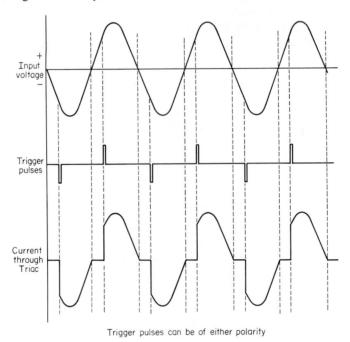

Trigger pulses can be of either polarity

Fig. 10.4. ac power control using a Triac.

end of the negative half cycle. By varying the phase of the trigger pulses, the amount of ac current flow can be controlled.

TRANSDUCERS

A *transducer* is any device which converts one form of energy into another form. Everyday examples are a microphone which converts acoustical energy into electrical energy, a loudspeaker (electrical energy into accoustical energy), a photocell (light energy into electrical energy), and a phonograph cartridge (mechanical energy to electrical energy). With electronic control, we are normally concerned with those transducers which convert energy of any form into an electrical output signal which can be used as an input to the electronic control system. Such transducers give an electrical signal proportional to the change of the physical quantity involved. Common quantities for which transducers are available include temperature, pressure, velocity, strain, vibration, shock, acceleration, displacement, force, tension, flow, and many others. Thus, with industrial equipment for practically any purpose it is possible to derive an electrical signal which can be used for measurement or control purposes. Mention of a few of the more common transducers will give some appreciation of what can be done.

A *thermocouple* consists of two dissimilar metals welded together at one end. When this welded junction is heated, a voltage is developed across the free ends. This voltage is proportional to the temperature difference between the welded junction and the free ends. The metals are usually in wire form; typical combinations are chromel-alumel for temperatures up to 1600°F and platinum/platinum–rhodium alloy for temperatures up to 2800°F. Thermocouples are used for the measurement of temperature and for the electronic control of furnaces and ovens.

Photoelectric transducers (photocells) are units which give an electric signal output when exposed to light. They can be used directly to measure and control the level of illumination or for counting items passing between the photocell and a light source. They can also be used to give indication of position and vibration. Displacement of an opaque vane interposed between a lamp and photocell will vary the light falling on the photocell and give dc output which varies linearly with the displacement. Longitudinal vibration of the vane will give an ac output from the photocell.

There are several types of *magnetic transducers.* One group, which operate by the relative motion between a magnet and a coil of wire, are called electronic tachometers. One of these is essentially a small dc generator which is driven by the rotating shaft to be controlled and gives a dc output proportional to the shaft speed. In another type, a small permanent magnet is attached to the shaft and a coil is positioned close to the shaft. When the shaft rotates, each time the magnet passes the coil, a voltage pulse is generated in the coil, and so the output is a train of pulses, the repetition frequency of which is proportional to the shaft speed. Both of these tachometers can be used for motor speed control.

A different type of magnetic transducer is based on the Hall effect. It consists of a wafer of semiconductor, along which a current flows. If a magnetic field is applied perpendicular to the wafer, a voltage difference, proportional to the magnetic field, is produced across the sides of the wafer.

Resistive transducers are units in which the ohmic resistance of an element is changed by relative elongation or strain. The most common of these is the strain gauge, which consists of a fine wire filament or thin-film stripe formed into a grid and attached to the item of which the strain is to be measured. Any strain on the item is also applied to the gauge and causes the wire to stretch, resulting in an increase in electrical resistance which can be detected by including the gauge as one arm of a resistance bridge circuit. More recently a thin wafer of silicon has been used in the same way with good operation. Strain-gauge transducers are used to give electrical signals proportional to the deformation or vibration of mechanical elements.

SOLID-STATE TRANSDUCERS

A very significant point is that most transducer energy conversions can be effected using solid-state techniques. It is possible to facilitate semiconductor structures which are electrically sensitive to changes in heat, light, deformation, and magnetic field. Such transducers will ensure good compatibility with integrated circuits and in some cases may be actually fabricated together with comparator, amplifier, and logic circuits in one integrated-circuit wafer. The feasibility of such assemblies has already been demonstrated, and when they become available as production items, they will add considerably to the scope and economics of electronic control.

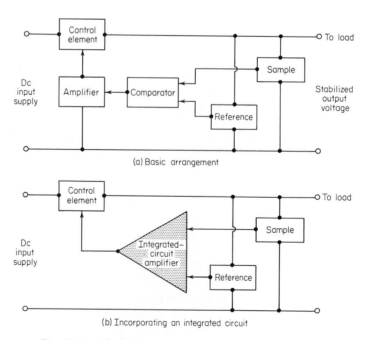

Fig. 10.5. Block diagrams of series voltage stabilizers.

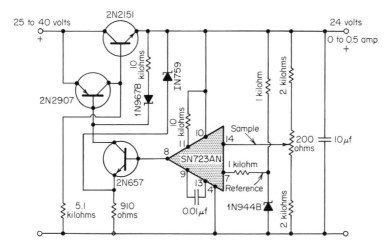

Fig. 10.6.   24-volt, 0.5-ampere voltage stabilizer incorporating an SN723AN linear integrated-circuit amplifier.

SELECTED EXAMPLES OF INDUSTRIAL ELECTRONIC CONTROL

To illustrate the part that integrated circuits can play in the field of industrial electronic control, several examples will be described.   Relatively simple applications using only small numbers of integrated circuits have been chosen so that the general operation can readily be appreciated.

Voltage Stabilization.   A requirement common to all electronic systems is a stable dc supply.   Voltage-stabilizer units have been established for some time as standard items of equipment using discrete components.   The block diagram of a conventional series voltage stabilizer is shown in Fig. 10.5a.   A sample of the output voltage is compared with a constant reference voltage.   When the output voltage is at the stabilized value, the sample voltage is equal to the reference voltage.   If it is different, the difference voltage is amplified and applied to the control element which adjusts to correct the output voltage back to the stabilized value.

A linear integrated circuit can conveniently be used for the functions of comparator and amplifier as indicated in Fig. 10.5b.

An example of a practical circuit incorporating an SN723AN integrated circuit (a three-stage medium-gain differential amplifier, generally similar to the SN72702N described in Lesson 7) is shown in Fig. 10.6.   Discrete power transistors are used in the control element.   The sample voltage is obtained from a resistance chain across the output, and the reference voltage is taken across a silicon zener diode D.   The integrated circuit compares these voltages and amplifies the difference to give an output which drives the control element.

The use of the integrated circuit results in a simpler assembly with better stabilization at a similar cost.   The circuit described will stabilize the output voltage of 24 volts to 0.01 percent, with the input voltage varying from 25 to 40 volts and the load current varying from 0 to 0.5 ampere.

Phase-controlled Trigger-pulse Generator. As indicated above, to control power with SCRs and Triacs, it is necessary to vary the phase of the gate trigger pulses. With feedback control systems using integrated circuits, we usually generate a control signal in the form of a dc voltage, the amplitude of which is proportional to the amount of control required. In order to use this varying dc voltage for control purposes, we must use it to produce a corresponding variation in phase of a trigger pulse. A convenient way of doing this is to compare it with a linear sawtooth voltage which is synchronized to the half cycles of the ac supply line. At the instant that the dc voltage equals the rising sawtooth voltage, a trigger pulse is generated. As illustrated in Fig. 10.7*a*, if the dc voltage is low, the equality will occur early in the half cycle and the trigger pulse will be correspondingly early. If the voltage is high, the equality will occur later, and the trigger pulse will be generated more to the end of the half cycle.

The circuit is suitable for fabrication as two integrated circuits, as shown in the block diagram of Fig. 10.7*b*. The comparator is a differential input circuit with the dc voltage and the sawtooth voltage as the two inputs. The pulse-generator unit contains the sawtooth charge and discharge circuits and the trigger-pulse generator circuit. The pulses are obtained by the discharge of capacitor C_2 through the primary of the pulse transformer.

This phase-controlled pulse generator can be used as a standard unit for a variety of electronic control applications as will be seen later.

Temperature Control. A basic system for temperature control of ovens and furnaces is shown in Fig. 10.8. A thermocouple located in the hot zone produces a voltage which unbalances a resistance bridge. The imbalance is detected and amplified by a differential integrated-circuit amplifier, and the dc output voltage is fed to a phase-controlled pulse generator as described above, which gives pulses to trigger the Triac and so control the power in the heater winding. If the oven temperature is too high, the excess thermocouple voltage results in a higher dc voltage at the output of the integrated circuit which delays the trigger pulses so that the heater power is reduced, thus bringing the temperature down to the required level.

The output voltage from the thermocouple is only of the order of millivolts, and so to give sensitive operation the system must respond to fractions of a millivolt change. This requires a high-gain integrated-circuit differential amplifier such as the SN72709N. (This has a similar gain to the SN725N described in Lesson 7.)

Illumination Control. For control of illumination, a very similar arrangement to that described for temperature control can be used, as shown in Fig. 10.9. The level of illumination is sensed by a cadmium sulphide or silicon photoconductive cell which is connected as one arm of a resistance bridge. Excess illumination lowers the resistance of the cell and results in a positive control signal, which is fed through a circuit identical to that described above, to delay the trigger pulses and reduce the power fed to the lamp.

Speed Control. Speed control of motors can be effected using either linear or digital systems. For a linear system, a dc tachometer is used to sense the speed; this gives a dc voltage output which is proportional to speed. The output from the tachometer is fed to a medium-gain integrated-circuit differential amplifier and compared with a reference voltage, as shown in Fig. 10.10*a*. Since the output level

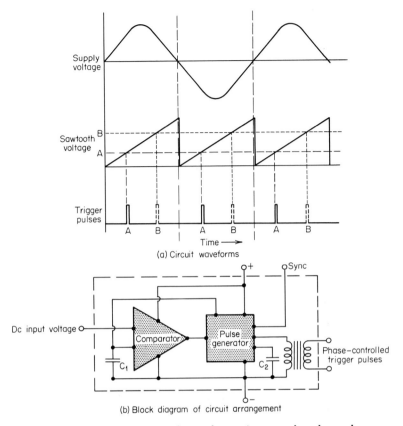

(a) Circuit waveforms

(b) Block diagram of circuit arrangement

Fig. 10.7.   Converting a dc voltage change into a pulse-phase change.

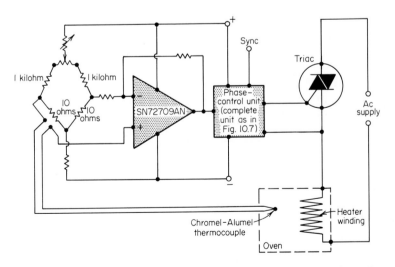

Fig. 10.8.   Temperature control system incorporating integrated circuits.

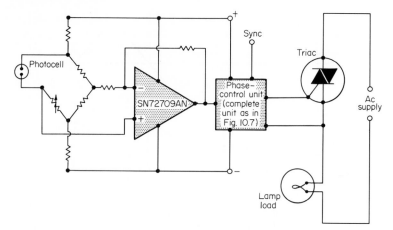

Fig. 10.9. Illumination control system.

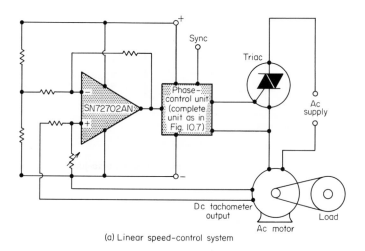

(a) Linear speed-control system

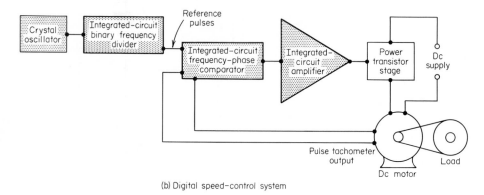

(b) Digital speed-control system

Fig. 10.10. Motor speed control systems.

from this type of tachometer can be relatively high, a medium-gain amplifier such as the SN72702N with negative feedback will be satisfactory. Again the output from the integrated circuit is fed to a phase-controlled pulse generator as described above, the resulting trigger pulses controlling the Triac in the motor-drive circuit.

For some applications, very precise control of speed is required, down to the order of 0.1 percent. In such cases, digital systems have proved to be superior to the linear approach. One method of using digital control with a dc motor is illustrated in the block diagram of Fig. 10.10b. The motor speed is monitored by a pulse-type tachometer which gives an output consisting of a train of pulses, the repetition frequency of which is proportional to motor speed. This train of pulses is compared, both in frequency and phase, with a reference train of pulses generated by a crystal oscillator and binary frequency divider. The output from the comparator circuit is amplified and fed to a power transistor stage which drives the dc motor. The frequency divider and frequency-phase comparator both use integrated circuits to advantage, about 10 standard digital circuits are required.

Solid-state Inverters. It is often required to operate ac equipment when only a dc supply is available. Early methods of doing this were to use a dc motor—ac alternator combination or a vibrating relay followed by a transformer. It is now possible to generate three-phase, ac supplies from a dc source, with closely controlled frequency and amplitude, using solid-state circuits. Integrated circuits play an important role in such equipment, called solid state inverters.

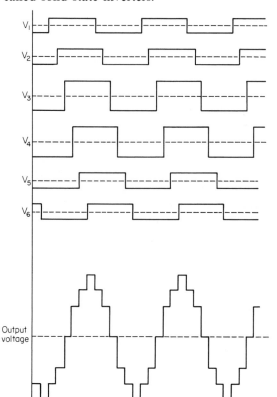

Fig. 10.11. Summation of square waves to give a step approximation to a sine wave.

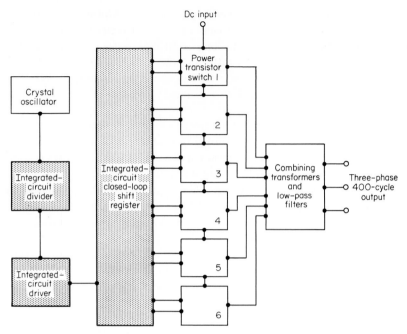

Fig. 10.12.   Block diagram of a solid-state inverter.

One interesting method used for a static inverter (i.e., an inverter with no moving parts) is to synthesize the sine wave output by a summation of several square wave pulses to give a step approximation as shown in Fig. 10.11.   A block diagram of a complete three-phase inverter is shown in Fig. 10.12.   A crystal oscillator generates a high-frequency signal of the order of 1 or 2 Mhz.   This is first divided down by an integrated-circuit ripple counter to 4,000 hz, and the output fed to an integrated-circuit closed-loop shift register, which produces 12 square-wave 400-hz signal voltages, with 30° phase separation.   These voltages drive six power-transistor switching stages in phase rotation, and the square waves at the outputs are combined to give a three-phase output which is passed through low-pass filters to smooth out the steps in the waveform.

If close control of frequency is not required, a simple, free-running multivibrator at 4,000 hz can be used instead of the oscillator and divider chain.

GLOSSARY

gate electrode   A control electrode to which a trigger pulse is applied.

photocell   A device of which the electrical characteristics are modified when it is exposed to light.

silicon controlled rectifier (SCR)   A silicon rectifier with a p-n-p-n structure, of which the forward characteristic can be triggered from the blocking state to the conducting state by a current pulse fed to a gate electrode.

strain gauge   A resistive transducer which gives an electric output signal proportional to its deformation under strain.

tachometer A transducer which gives an electric output signal proportional to the rotational speed of a shaft.

thermocouple Two dissimilar metals welded together at one end. When this welded junction is heated relative to the free ends, a voltage is developed across the free ends.

thyristor Any semiconductor device that operates on the p-n-p-n regenerative principle.

transducer A device which converts one form of energy into another form.

Triac A device similar to two silicon controlled rectifiers connected in reverse parallel. Both the forward and reverse characteristics can be triggered from the blocking to the conducting state.

REVIEW

For each of the numbered statements below, select the one of the items lettered *a, b, c,* or *d* that correctly completes the statement.

10.1. In industrial electronic control
    a. The use of integrated circuits is not important.
    b. Integrated circuits can be used with advantage in feedback control circuits.
    c. ac power systems are always used.
    d. We are only concerned in establishing stable control with constant load conditions.

10.2. The silicon controlled rectifier
    a. Has two electrodes.
    b. Can be triggered to pass current in both directions.
    c. Always blocks in the reverse direction.
    d. Controls the output power by passing current during the full forward half cycle and varying its amplitude.

10.3. A Triac
    a. Is equivalent to two SCRs connected in reverse parallel.
    b. Is a three-phase device.
    c. Is a three-region n-p-n device.
    d. Always conducts in the forward direction.

10.4. When using a Triac for ac power control
    a. It is always necessary to use positive trigger pulses.
    b. It is normal to trigger during the positive half cycles only.
    c. The trigger pulses can have either polarity.
    d. Reverse current flows even with no trigger pulse.

10.5. With phase control of power using SCRs or Triacs
    a. The phase of the output current is varied.
    b. The phase of the output voltage is varied.
    c. The phase of the input voltage is varied.
    d. The phase of the trigger voltage is varied.

10.6. A transducer
    a. Is a device which converts energy of one form into another form.
    b. Is another name for a transformer.
    c. Changes ac electrical energy into dc.
    d. Operates in the same way as a transistor.

10.7. In a dc voltage-stabilization circuit,
    a. A transducer is used to sample the output voltage.
    b. A linear integrated-circuit amplifier can be used to amplify the feedback signal.

    c. Phase control is normally used.

    d. A Triac is used as the main control element.

10.8. In temperature control systems, integrated circuits

    a. Are used instead of thermocouples.

    b. Are used as the main power control element.

    c. Need only have low gain.

    d. Compare the output of a thermocouple with a reference and amplify the difference signal.

10.9. To control the speed of electric motors

    a. Only linear systems can be used.

    b. Digital systems using digital integrated circuits are superior for precise control.

    c. Resistance transducers are used.

    d. Very-high-gain integrated-circuit amplifiers are essential in the feedback circuit.

10.10. A static inverter

    a. Is a stationary transformer.

    b. Is an equipment for reversing static electricity.

    c. Is an equipment which converts dc power to ac power with no moving parts.

    d. Cannot be fabricated using solid-state devices.

Answers to Review

1.1 *c*	**1.2** *b*	**1.3** *d*	**1.4** *a*	**1.5** *c*	**1.6** *c*	**1.7** *d*	**1.8** *b*	**1.9** *c*	**1.10** *a*
2.1 *c*	**2.2** *b*	**2.3** *d*	**2.4** *d*	**2.5** *c*	**2.6** *a*	**2.7** *b*	**2.8** *c*	**2.9** *a*	**2.10** *c*
3.1 *c*	**3.2** *a*	**3.3** *c*	**3.4** *b*	**3.5** *d*	**3.6** *c*	**3.7** *d*	**3.8** *b*	**3.9** *c*	**3.10** *a*
4.1 *c*	**4.2** *c*	**4.3** *d*	**4.4** *a*	**4.5** *b*	**4.6** *d*	**4.7** *c*	**4.8** *b*	**4.9** *b*	**4.10** *c*
5.1 *b*	**5.2** *d*	**5.3** *c*	**5.4** *b*	**5.5** *c*	**5.6** *c*	**5.7** *b*	**5.8** *a*	**5.9** *d*	**5.10** *c*
6.1 *c*	**6.2** *a*	**6.3** *b*	**6.4** *c*	**6.5** *d*	**6.6** *c*	**6.7** *d*	**6.8** *b*	**6.9** *c*	**6.10** *b*
7.1 *b*	**7.2** *c*	**7.3** *d*	**7.4** *b*	**7.5** *d*	**7.6** *a*	**7.7** *a*	**7.8** *c*	**7.9** *d*	**7.10** *c*
8.1 *d*	**8.2** *b*	**8.3** *c*	**8.4** *d*	**8.5** *a*	**8.6** *c*	**8.7** *a*	**8.8** *b*	**8.9** *c*	**8.10** *b*
9.1 *c*	**9.2** *d*	**9.3** *d*	**9.4** *a*	**9.5** *c*	**9.6** *b*	**9.7** *d*	**9.8** *b*	**9.9** *a*	**9.10** *c*
10.1 *b*	**10.2** *c*	**10.3** *a*	**10.4** *c*	**10.5** *d*	**10.6** *a*	**10.7** *b*	**10.8** *d*	**10.9** *b*	**10.10** *c*

Index

53657

DATE DUE